Fisheries Biology

A Study in

Population Dynamics

Fisheries Biology
A Study in
Population Dynamics

Second Edition

D. H. Cushing

THE UNIVERSITY OF WISCONSIN PRESS

Published 1981

The University of Wisconsin Press
114 North Murray Street
Madison, Wisconsin 53715

The University of Wisconsin Press, Ltd.
1 Gower Street
London WC1E 6HA, England

Second edition 1981

Printed in the United States of America

For LC CIP information see the colophon

ISBN 0-299-08110-9

Contents

Illustrations

Tables

Preface to the Second Edition

In the spring of 1963, Professor A. D. Hasler asked me to give a series of lectures on fisheries biology to his graduate students in the Department of Zoology of the University of Wisconsin at Madison. The present text has been developed from those lectures, and is offered as a basis on which to build management and conservation principles for commercial fisheries.

Material is taken from the literature published on the results of fisheries research, some freshwater, but mainly marine, throughout the world. Because I live on the east coast of England, the literature I have used principally concerns research and problems dealing with the fisheries of northern Europe. Much of the important scientific literature is buried in the papers of more or less inaccessible journals. The basic concepts are those of E. S. Russell and Michael Graham, both former directors of the Fisheries Laboratory, Lowestoft, and of W. E. Ricker of the Fisheries Research Board of Canada. These ideas were put in analytic form by R. J. H. Beverton and S. J. Holt in their book *On the Dynamics of Exploited Fish Populations* (1957); in addition, their analyses made full use of growth and mortality data in yearly increments. To this list we should add the use of cohort analysis developed by John Gulland, which now forms the basis of fish population studies wherever fish have been aged for a decade or so.

Fish taxonomists are few and overworked, and up-to-date scientific names covering the scope of the present volume are unavailable in any one publication. Where pertinent geographically, the nomenclature I have used follows that found in *Plymouth Marine Fauna* (Marine Biological Ass. of the United Kingdom, Plymouth; 2nd ed., 1931; 371 p.), *A List of Common and Scientific Names of the Better Known Fishes of the United States and Canada* (Amer. Fish. Soc., Washington, D.C.; Spec. Pub. No. 1, 3rd ed., 1970; 150 p.), G. V. Nikolsky, *Special Ichthyology* (published for the National Science

Foundation and the Smithsonian Institution by the Israel Program for Scientific Translations, Jerusalem, 1961; 538 p.), and *Bulletin Statistique des Pêches Maritimes Cons. Perm. Intern. Explor. Mer*, 49 (1966), p. 72–80. The notations used are based on those of Beverton and Holt, as well as those in "A Standard Terminology and Notation for Fishery Dynamics," by S. J. Holt, J. A. Gulland, C. Taylor, and S. Kurita (1959, *J. Cons. Intern. Explor. Mer*, 24:239–42), accepted by the International Council for the Exploration of the Sea in Bergen, 1957.

I am grateful to my colleagues J. A. Gulland, F. R. Harden Jones, and A. C. Burd for reading the manuscript of the first edition during the spring of 1964, and to Harold Jenner for the drafting work that most of the figures required. I am also grateful to H. A. Cole, Director of the Fisheries Laboratory at Lowestoft, not only for reading the text, but for allowing me to travel to Madison. To Professor Hasler I am indebted for having invited me there. The trip was made possible by a training grant from the United States Public Health Service.

The Cartographic Laboratory at the University of Wisconsin-Madison assisted in making modifications of the figures needed in the second edition.

This second edition comprises some of the researches published since 1967, when the first edition was completed. The present text is a little more extensive and the arrangement of chapters has been changed. This is because much more experimental work on the growth of fishes has been published, and the subject has become part of the preliminary biology rather than an adjunct to population analysis. Further, our knowledge of the dependence of recruitment on parent stock has increased considerably, and a larger chapter on that subject was needed. I am grateful to my colleagues F. R. Harden Jones, Alan Jamieson, John Pope, David Garrod, and Joe Horwood, who read parts of the second edition. I am also grateful to Arthur Lee, Director of the Fisheries Laboratory at Lowestoft, for reading the whole text.

D. H. C.

Fisheries Laboratory
Lowestoft, Suffolk

Fisheries Biology

A Study in

Population Dynamics

1

The Scope of Fisheries Biology

Between 1955 and 1965, the world's catch of fish in freshwater and in the sea doubled. Before the collapse of the Peruvian anchoveta (*Engraulis ringens* Jenyns) fishery in 1972, the total catch was just over 70 million tons. Some stocks, such as the sardine-like fishes off southern Arabia, remain under-exploited, but they are few and far apart. Others, like the tuna species in all the subtropical oceans, are exploited at about the right rate. But most of the bottom-living fish stocks in the North Atlantic and North Pacific are over-fished, and the herring fisheries in the Northeast Atlantic have been nearly extinguished. Further, Antarctic blue whales (*Sibbaldus musculus* [Linnaeus]) have been fished nearly to extinction. It is unlikely that they will recover for many, many years; an annual production of more than 1 million tons has been lost. This book describes methods used in the study of fisheries biology by which stocks may be measured, conserved, and properly exploited.

Until recent years, stocks of fish in the open sea were the common property of all nations, and fishermen found that the more heavily they fished the more their catches declined. As a consequence, most countries established laboratories where scientists have been able to study the biology of the fish stocks, their population dynamics, and the means of sharing them internationally. As the third United Nations Law of the Sea Conference reaches its conclusion, most maritime nations have declared 200-mile limits within which they claim prior rights to exploit fish stocks among other resources. Where catches were shared out by the international fisheries commissions, they are now to be exploited predominantly by the coastal state or states. Historically we have moved from international control to coastal state control; it is likely, however, that international scientific organizations will remain because fish are not usually restricted to the waters of one coastal state. Very large quantities of information have been collected, and some fish stocks are now the most fully studied of all wild animal populations.

There are two branches of fisheries biology. One is the study of the natural history of the stocks, and is concerned with how the fish spawn, grow, and feed. Its primary purpose is to delimit the stocks, or the unit populations, to establish how they grow and survive at all stages of their life histories and how they relate to trophic levels above and below their own. The other branch is the study of the dynamics of such populations — the rates at which fish grow, reproduce, and die. Most fisheries biology consists of the description of individual fisheries in terms of the fish stocks and their dynamics. The two branches become linked in any study of the stabilization of populations or in one of multispecies fisheries.

Fish Population Studies

In a very simple form of population dynamics (Fig. 1), the presence of senile fish indicates very little fishing pressure. The figure represents two *Tilapia* (*Tilapia esculenta* Graham) caught in Victoria Nyanza in East Africa by Michael Graham (Graham, 1958). The top one is a fish of moderate age in good condition; the bottom one appears to be a senile animal, with its heavy jaw and gill cover and its large pectoral fin. The first was taken from a well-fished population in the Kavirondo Gulf of Victoria Nyanza, and the second from a very lightly fished population in the Emin Pasha Gulf on the same lake. Most of the fish in the catch in the Emin Pasha Gulf were old ones. There were very few old ones in the much larger catches from the Kavirondo Gulf. Moderate or heavy fishing increased the chance of death and eliminated most of the senile fish. Senile plaice (*Pleuronectes platessa* Linnaeus) were noticed by Garstang (1900–1903) in the North Sea and by Petersen (1894) in the Skagerrak, and senile American plaice (*Hippoglossoides platessoides* [Fabricius]) were recorded by Templeman and Andrews (1956) off New-foundland. But little is known of senile fishes in the great populations of commercial fishes, most of which are heavily exploited.

Market Statistics. A fair fraction of the fish population is counted and weighed on the fish quays and markets. A complex system has been developed to sample the catch on the quay and the stock in the sea. From each ship, details of catch, position, and number of fishing hours are recorded on the fish quays. The time spent fishing is called the fishing effort. The catch per unit of fishing effort (i.e., cwt/100 hr trawling, number of tuna per 100 hooks, etc.) is a proper index of stock (Ricker, 1940; Beverton and Holt, 1957). Consequently, charts can be made of catch per effort (or stock density) of different species on different grounds month by month. The charts of stock densities in 5° squares of various tuna species in the subtropical regions of the world ocean, issued by the Far Seas Laboratory in Shimizu, Japan, are the most extensive and detailed distribution studies of any wild animal. British catches

Figure 1. An elementary form of population analysis. The top picture is of a middle-aged *Tilapia* caught from a moderately fished stock in Victoria Nyanza. The lower picture is of a senile fish from a relatively unfished stock. Senile fish are easily recognizable and readily caught; they are not found much in a moderately fished stock. The presence of senile fish thus indicates absence of fishing. From Graham, 1958.

of many species have come from the North Sea, Faeroe Islands, Iceland, Barents Sea, and West Greenland, and in the 1960s British fishermen returned to the Grand Banks off Newfoundland. In British waters, catches have been recorded in statistical rectangles of 30 INM × 30 INM (International Nautical Miles) since the 1920s. Off the eastern seaboard of the United States and Canada, analogous records are supported by groundfish surveys.

Not only are records made of catch, position, and time spent fishing, but fish are also measured for length at the market. With its nose rammed against the stub of a board, each fish is smoothed down to the tail, and measured to the end of the spread tail fin. The measurers take a fish box, the weight of which is known, and count and record the lengths of fish in it. Hence, a very complete picture of the length-frequency distribution of the catch is constructed for each area, each day, each week, or each month. It is quite simple to express such distributions in catch per unit of effort or stock density (e.g., no. cwt of 60-cm cod [*Gadus morhua* Linnaeus] per 100 hr fishing from Bear Island, between Norway and Spitsbergen, in June). In 1961, 750,000 fish of various species were measured in this way in English ports (Great Britain. . . . 1962. Fish stock record, 1961). Good estimates are thus available of the length distribution of the catch and also of catch per unit of effort — and hence, the stock. In recent years, British catches have been recorded by 3, 6, 9, and 12 mile belts in an effort to study the inshore fishes. Under the auspices of the International Council for the Exploration of the Sea, roundfish in the North Sea are now sampled by areas rather than by port.

Growth Rates and Death Rates. Because the ages of fish can be readily measured, it is possible to estimate the growth rates and death rates of fish populations. Otoliths are collected from most fish, but scales are taken from herring (*Clupea harengus harengus* Linnaeus) and salmon (*Oncorhynchus* sp. and *Salmo* sp.). Figure 2 shows an otolith from a plaice and a scale from a herring. The age of each fish is readily determined from the annual rings found on both otolith (Hickling, 1931; Rollefsen, 1934) and scale (Dahl, 1907; Lea, 1929). Recently (Brothers, Mathews, and Lasker, 1976) it has been discovered that there are daily rings on the otoliths of the Pacific hake (*Merluccius productus* [Ayres]) and on the larval otoliths of the northern anchovy (*Engraulis mordax* Girard), and so they also can be aged. In the future, studies of the growth rates and death rates of fish larvae may well lay the foundations of our understanding of the stabilization of fish populations.

Age-and-length correlation tables are constructed from samples of scales and otoliths; the tables are then used to convert the large numbers of length measurements into estimates of age (Fridriksson, 1934). For example, during the 1950s 400 fish a week were aged in the East Anglian herring fishery, off the east coast of England, but up to 10,000 fish a week were measured in length. (Today the statistical bases of such correlation tables are more fully understood; now we would measure fewer fish and collect more scales and otoliths [see Tanaka, 1960; Kimura, 1977; Brander, 1974, for an account of the statistical structure of age-length keys].) Thus we are able to express both the catch and the stock in terms of age, e.g., as so many cwt of four-year-old fish caught or as per 100 hr fishing in the stock. Because fish can be easily aged and this information can be carried right through the sampling system, the ages of fish populations are very much better known than those of other wild species. Lengths for an age class, based on the catches, can be compared from year to year to yield growth rates. Within a fully recruited year class (i.e., a brood of fish born in a given year), the decline in stock density from

Figure 2. An otolith from a plaice (*top*) and a scale from a herring (*bottom*). The rings on both are annual rings, so the ages of both species can be readily determined.

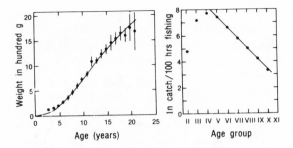

Figure 3. Left. The growth in weight of the plaice in years of life. The estimates are taken from market measurements and are averaged for the period 1929–38. *Right.* The mortality of plaice between the ages of five and ten years. The estimates are taken from the system of market measurements and statistics of catch per unit of effort. They were averaged for the period 1929–38. Both figures adapted from Beverton and Holt, 1957.

year to year is due to mortality; the ratio of stock densities can thus be used to estimate mortality. Figure 3 (*left*) shows the increase in weight for various ages of plaice based on market measurements, and Figure 3 (*right*) shows the average mortality of plaice for a period of ten years, based on estimates of stock density. Both estimates are averaged from a large quantity of information collected from the market, and they are the bones of fisheries research.

Independent estimates of stock. Because estimates of stock density can become biased (see Chapter 5), independent measures of the stock in the sea are sometimes needed. Where long-time series of catches-at-age are available, good estimates of stock by age and year can be made for past years with cohort analysis (see Chapter 5). But stock estimates of a present year are also sometimes needed for the establishment of quotas. Then independent estimates are needed, by groundfish survey, by acoustic survey, and by egg survey. As an example we examine an egg survey. Most commercially fished species, except herring, lay eggs that float in midwater. Such eggs can be readily caught by silk or nylon plankton nets and counted, as in the case of pilchard eggs (*Sardina pilchardus* [Walbaum]) in the English Channel (Fig. 4). Five cruises were made in April, May, June, July, and August 1950, extending over the entire Channel (Cushing, 1957). The area was split into statistical rectangles, with a number of stations in each rectangle. The number of eggs produced in the sum of rectangles at a point in time (e.g., the midpoint of a cruise) is given as

$$\Sigma\ (OAQ),$$

where O is the average number of eggs caught beneath one square meter in an early developmental stage in one rectangle,

A is the area of that rectangle in m², and

Q is the temperature factor, the number of days in the month divided by the temperature coefficient of egg development.

Each point on the curve represents the sum for the whole Channel (Fig. 4); the whole curve represents the total production of stage I eggs (hatched eggs before the appearance of the neural crest) through a rather extensive spawning period. The eggs take 40–72 hours to hatch in the Channel, depending on the temperature. Because it is a rather short period, errors in temperature conversion are likely to be small ones. The standard deviation for one cruise is

$$\overline{Q} \ \sqrt{\ \Sigma \ (O^2 A^2 s^2 / k),}$$

where s^2 is the variance of the estimate of numbers in each rectangle,

k is the number of stations in the rectangle, and

$\overline{Q}$ is the average temperature factor for the whole cruise; errors in temperature determination and in the estimate of the temperature coefficient have been ignored.

The number of adult pilchards in the English Channel is given by the number of eggs produced (Fig. 4), divided by twice the number of eggs per female. With a revised estimate of fecundity, Macer (1974) has shown that the quantity should be $5 . 10^9 \times 2.2, \times 0.45$. The error, or sampling variation, is a multiplicative one because the distribution is a log normal one. English (1964) has pointed out that the error between cruises is high and that the cost in cash of reducing it to manageable proportions may even be prohibitive. However, Lockwood (1978) has shown recently that the errors in an egg survey in one particular case (mackerel [*Scomber scombrus* Linnaeus] off the western coasts of the British Isles) were low. In principle, this estimate of population numbers can be used as a check against the stock-density estimates made with catches per unit effort.

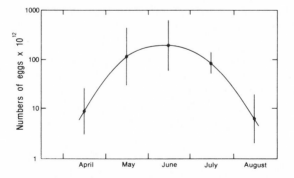

Figure 4. The production of pilchard eggs in the English Channel. The vertical lines indicate the error on each estimate of production. Adapted from Cushing, 1957.

The Theory of Fisheries Dynamics. The fourth characteristic of fisheries biology is common to all population dynamics; it is the mathematical theory. As in many branches of science today, it involves the making of models. Model-making is a form of conceptual experiment and is probably the only way of unraveling complex situations where many things happen at once. Today, ecologists and oceanographers use models extensively to describe and summarize interlocking events, much as fisheries biologists have done since the 1920s and 1930s. The theory of fish populations has been developed by Baranov (1918), Graham (1935), Beverton and Holt (1957), and Ricker (1958a); Ricker summarizes and refers to a number of his own papers published between 1940 and 1958. But the simplest formulation of the theory of fishing is Russell's (1931). His equation is

$$P_2 = P_1 + G + R'' - Z', \tag{1}$$

where P_2 is stock in weight at time t_2, in the second year,
P_1 is stock in weight at time t_1, in the first year,
G is the increment in weight due to growth between t_1 and t_2,
R'' is the increment in weight due to recruitment between t_1 and t_2, and
Z' is the decrement in weight due to mortality between t_1 and t_2.
If the stock is to remain in a steady state, the gains from growth and recruitment must be balanced by losses due to mortality.

A more complex derivation of this equation appears in Chapter 5, where stock is expressed in terms of numbers or weight, and where the mathematics and notation of Beverton and Holt (1957) are used. These authors use the description of growth and mortality with reference to age as an essential part of the central equations in an analytic manner. This is the major advance that they have made beyond those of the other theoreticians. However, Graham's modification of the logistic curve has been developed by Schaefer (1954, 1957), and subsequent use of this simple model has some advantages where fish cannot or have not been aged and where fisheries comprise many species mixed together on the grounds.

Above, then, are the four branches of population dynamics used in fisheries research. The first is the sampling of the catch on the quay and of the stock in the sea. The second is the conversion of length measurements made at the market into estimates of age in units of stock density or in units of catch. The third branch is the creation of independent measures of stock density. The fourth is the development of models in mathematical terms.

The Biology of the Stocks

The powerful theoretical weapons developed for analyzing the changes in fish populations are of little use if the populations cannot be delimited. Sometimes the population is isolated by deep water (like Faeroe cod) and the methods can be used directly. Where populations mix, extensive biological analysis is required to delimit them. Three types of biological work are needed for this purpose: the study of migration; the study of the unit stock in itself, that is, subpopulation characteristics of phenotypic or genotypic origin; and the study of growth and recruitment, where characteristics often differ between stocks or subpopulations.

Migration. The study of migration is fundamental to fisheries research. The fish appear to make use of the current structures for purposes of spawning, feeding, and the positions of their nurseries (Meek, 1916). Eggs and larvae drift away from spawning ground to nursery ground; the adults migrate back to the spawning ground. This ensures that the larvae drift to a nursery ground close to the feeding ground. In this sense fish stocks may be considered as being contained within persistent oceanographic structures. For example, albacore (*Thunnus alalunga* Bonnaterre) that have been tagged off the California coast migrate to fishing grounds close to Japan (Otsu, 1960, see Fig. 31, Chapter 3). Most journeys last about a year, but some are less. On a straight course, this would be fast — but when the trip is made by traveling at random around the North Pacific gyre, it becomes a very fast migration. It is possible that the albacore live constantly in the gyre, for the larger fish are caught in the North Equatorial current and spawn in a broad area east of the Philippines, but the immatures live in the north, as indicated by the monthly progress of catches along the Kuroshio and eastward into the Kuroshio extension. Thus, this particular stock might be considered as a body of fish living in the North Pacific gyre.

Similarly, the Norwegian herring drift in a circuit around the Norwegian Sea (Devold, 1963). As the stock spawning along the Norwegian coast near Bergen comprises the same fish that visit Jan Mayen and Iceland in the summer, the group ranging within a unit of water circulation might possibly be unified into a single stock. A single circuit may be formed from the Atlantic current flowing north off the Norwegian coast, the polar front off Jan Mayen, and the East Icelandic current which flows from that region toward the Faeroe Islands. How the herring migrate consistently within this structure of ocean currents is unknown.

Growth and Feeding. The biology of the North Sea herring is considerably influenced by the quality of the food, for their growth changes have been associated with changes in the quantity of the copepod *Calanus finmarchicus* Gunnerus in the sea (see Chapter 4). It is possible that a whole complex of fisheries originated from changes of growth, which in turn cause changes in recruitment, the pattern of recruitment, and hence in the disposition of fisheries. Hardy et al. (1936) started much of this work when they gave small

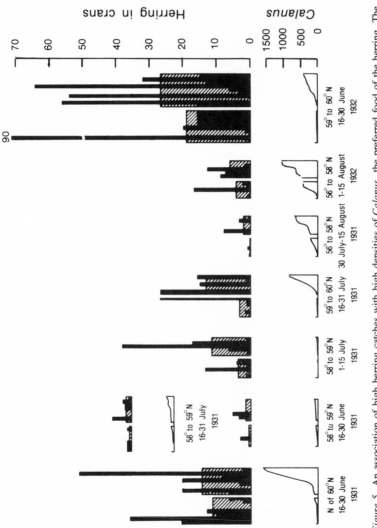

Figure 5. An association of high herring catches with high densities of *Calanus*, the preferred food of the herring. The catches in crans (c. 180 kg) for dense and thin aggregations of *Calanus* are shown as histograms, and the diagonal hatching indicates the average catches. The density of *Calanus*, estimated on the same night as the catch, is shown below the histograms as a thin line split at the median to separate "rich" and "poor" concentrations of *Calanus*. Adapted from Hardy et al., 1936.

plankton nets to the driftermen and taught them to recognize *Calanus*. The
nets, torpedo-like in shape and operation, were towed astern of the ship. In
one summer at least, high catches of herring were associated with high catches
of *Calanus* (Fig. 5). Because they aggregate at patches of *Calanus*, the herring grow more quickly, particularly if the food patches are extensive. Since
the growth of fish is directly modifiable by such environmental changes, this
field of study is important from the point of view of population change. It is
often assumed that a population is modified in quantity only at recruitment.
The changes in herring growth in the 1950s and early 1960s in the North Sea
account for an increase in weight of stock of nearly 20 percent. With
hindsight, it is possible that changes due to fishing on the North Sea herring
were masked by this increment in weight of stock — at least from a superficial
point of view.

Spawning. Two groups of spawning grounds are used by the southern North
Sea herring (Fig. 6): (a) Sandettié, Cap Blanc Nez, and Hinder grounds; (b)
Ailly, Creux St. Nicolas, and Vergoyer. Each is small, perhaps 2–3 km long
by 500 m wide. Each year the fish spawn in the same places. The exact
positions were discovered by trawlers using echo sounders; very dense areas
of echo traces were found only there and on the assembly grounds. Trawlers
did not work elsewhere between 1950 and 1963, except on one occasion in
1951 when they found a patch of spawning herring southeast of Beachy Head.
The timing of spawning is very regular. The fish appear near the Sandettié
Light Vessel during the first ten days of November (Ancellin and Nédelèc,
1959). On these grounds in November and December 1962, 150 large vessels
were working by day and perhaps 100 smaller pair trawlers by both day and
night.

Not all fish spawn on groups of restricted grounds, although herring may
well do so (Norwegian herring, Runnstrom, 1934; Firth of Clyde herring,
Stubbs and Lawrie, 1962; British Columbian herring, Harden Jones, 1968;
Hokkaido herring, Motoda and Hirano, 1963). Plaice spawn on broad patches
50–100 km across, as if each female was 80 m distant from the next (Beverton, 1962). The Arcto-Norwegian cod spawn in a restricted layer in the Vest
Fjord in northern Norway (Hjort, 1914; Bostrøm, 1955), but the Icelandic cod
spawn on extensive grounds, south and southwest of Iceland (Schmidt, 1909).
The albacore spawn in the North Equatorial Current between mid-ocean and
the Philippines (Matsumoto, 1966). The diverse spawning behavior probably
reflects differing needs in establishing migratory circuits and thence unit
stocks.

Fisheries biology must ultimately depend on the proper answer to the
biological problems here: (a) to understand the spawning mechanisms and
how the fish remain within the very narrow spawning grounds; (b) to un-

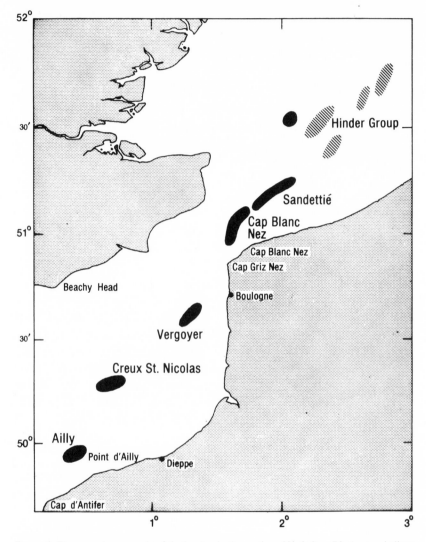

Figure 6. The spawning grounds of the herring in the southern North Sea. Black areas indicate spawning grounds; diagonally hatched areas are assembly grounds. Adapted from Ancellin and Nédelèc, 1959.

derstand how the herring and other fish can appear on the spawning grounds so regularly, when their cruising speeds are less than the maximum speeds of the tidal streams or currents. If the fish spawn regularly on the same restricted grounds, they may belong to the same spawning group, and such evidence can be used to establish the unity of a fish stock or subpopulation.

The Biology of a Fishery

Because fisheries biology is a combination of population dynamics and natural history, any fishery should be describable in terms of the biology of the stock and of its dynamics. This theme is illustrated by the Downs herring of the Southern Bight of the North Sea and the course of the fisheries that used to take place there in autumn and winter (Fig. 7). The Downs herring comprise a population of North Sea herring, the characteristics of which are noted below. South of 53°30′ N there is one group of herring that spawns on the grounds just described. North of this line, around the edges of the Dogger Bank, there were trawl fisheries on spawning herring from mid-September to mid-October. These exploited another stock — the Dogger stock, as opposed to the Downs stock which spawns in the Straits of Dover and in the eastern English Channel. The first fisheries for Downs herring were drift-net operations in five areas during October and November: (1) north of the Norfolk Banks, north of Lowestoft; (2) near Smith's Knoll Light Vessel, northeast of Lowestoft; (3) on the Brown Ridges, east of Lowestoft; (4) on the Schouwen ground, east-southeast of Lowestoft; and (5) on the Galloper ground, south-southeast of Lowestoft. These small subfisheries operated roughly in the order given as the herring move south. In November and December, trawl fisheries were found on the spawning grounds at Sandettié and Ailly. Then from December until March, pair trawlers worked close to the French, Dutch, and Belgian coasts on the spent herring (spawned) drifting north along the continental coast. The fisheries thus moved south as the herring moved south, and north as the spent fish moved north. The old Lowestoft spring herring fishery, which died out in the 1930s, was an essential link in the chain of migration. It started off the Dutch coast and moved from there to the Brown Ridges, the Norfolk Banks, and to the Silver Pit (just south of the Dogger Bank), between February and April. Some evidence from tagging experiments have shown that the herring did migrate, as did the fishery (Bolster, 1955).

The biological data in the East Anglian fishery can be summarized in two diagrams. Figure 8 shows the age sequence in stock density by weeks in the East Anglian fishery for 1935 (Hodgson, 1957). The young fish, those three, four, and five years old, first entered the fishery in October; the older fish, those six, seven, and eight years old, came in November. Because the younger fish came earlier and the older ones later, the age-sampling, in order to be adequate, had to be completed in short time intervals of three days or a

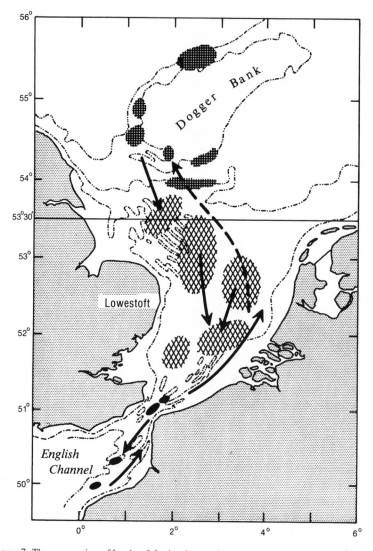

Figure 7. The progression of herring fisheries that used to take place in the Southern Bight of the North Sea. The vertically hatched areas represent the spawning fisheries around the Dogger Bank in September and October. The diamond hatching shows the positions of the drift-net fisheries in the Southern Bight in October and November. The black arrows pointing southward represent the southerly movement of the fish through the fisheries. The black patches indicate the trawl fisheries that used to take place on the spawning grounds in November and December. The long black arrow along the continental coast represents the course of the pair-trawler fishery that used to work for spent herring in December, January, and February (and in some years, later). The dashed arrow represents the course of the old Lowestoft spring drift-net fishery in the 1920s, which lasted from February to April. Depths (in ten and twenty fathoms) are indicated by the finer lines of dots and dashes.

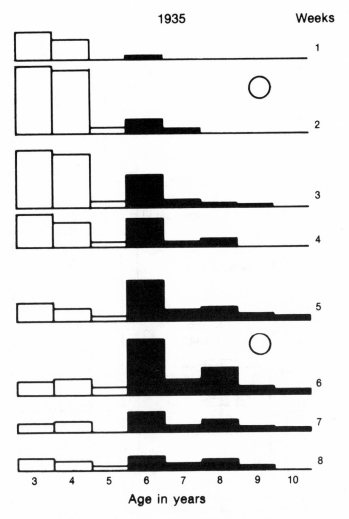

Figure 8. The age distributions (abscissa) in stock density (ordinate) in the East Anglian herring fishery for eight successive weeks in 1935, from early October (weeks 1–4) to late November (weeks 5–8). The three-, four-, and five-year-old fish (white columns) entered the fishery in November. The six-, seven-, and eight-year-old fish entered in November. The week of the full moon is shown by the open circles; the highest catches were made at or near the date of the full moon. Adapted from Hodgson, 1957.

18

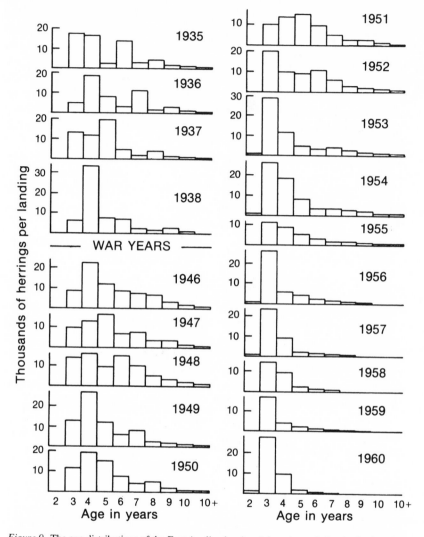

Figure 9. The age distributions of the East Anglian herring fishery in stock density from 1935 to 1960. Before 1952, the fish recruited partly at three years of age and partly at four; in 1952 and subsequently, recruitment was complete at three. Note the severe loss of older fish during the second half of the 1950s.

week. Figure 9 shows the total age distributions in numbers per drifter shot (or stock density) from 1935 to 1960. The same sort of picture was found in the samples from the trawl fishery based on Boulogne and in those from the spent herring fishery based on Ostend. There are two notable biological features — a change in the recruitment pattern in 1950–52, and the gradual loss of the older fish in the later 1950's, due to increasing mortality. Before 1950 the recruits came into the fishery at two time intervals, some at three years of age and others at four, a year later. By 1952 all recruits came in at three years of age; this situation remained until the fishery was extinguished in the middle sixties by recruitment overfishing.

Population accountants can manipulate the methods of population dynamics with little reference to the biology of the stocks. A study of a fishery in detail, which is linked to migratory behavior, growth metabolism, and spawning activity, illuminates the population dynamics. Only in this way can the methods of population dynamics be used properly and fully.

Conclusion

There are two branches of fisheries biology. The first concerns the dynamics of the stocks, the estimates of catch and stock density on the markets, the ageing of the fish in terms of stock density, the use of independent measures of abundance, and the use of mathematical models. The second branch involves the study of migration, growth, spawning, and recruitment in terms of the stock distribution. An extension of fisheries biology describes fisheries in terms of the biology of the stock and its dynamics.

The detailed biology is extensive in range and requires the basic physiology of metabolism (as applied to growth studies), of sensation (as applied to the study of migration), and of endocrines (as applied to the investigation of reproduction). To understand the oceanic environment, considerable knowledge of the physics of the ocean is needed, for example, how water masses move and how light penetrates the sea. Particular instrumental skills are needed, such as a knowledge of acoustics and electronics. The extensive range is also exemplified in the results of dynamic studies of the populations themselves. Very large quantities of data are available on distribution in time and space of the stock densities of many fish species. There are diverse methods of handling the material used in association with tagging techniques and other forms of population estimates.

In this book, three chapters deal with the preliminary biology needed for any stock analysis: the study of migration, the study of unit stocks or subpopulations, and the investigation of the growth and death of individual animals. The next two chapters describe the measures of abundance on which all stock analysis is based and the causes of mortality, natural and fishing, on which the management models are based. The seventh chapter is concerned with the current central problem of fisheries biology, which is the dependence of recruitment upon the magnitude of the parent stock. The models used in management are described in Chapter 8. The subsequent chapter examines the oceanic boundaries at which fish gather and the dramatic changes of fish populations in time. The final chapter, entitled "The Future of Fisheries Research," describes some areas of interest which might develop into valuable lines of research.

2
Migration

The study of migration is fundamental to fisheries biology because the migration circuit delimits the area in which the stock or subpopulation lives, thus ensuring the stock's reproductive isolation. The eggs of most marine fishes are about 1 mm in diameter and are transparent and pelagic; like them, their larvae drift with the tides and the ocean currents. From the spawning grounds the larvae may drift for considerable distances toward the nursery grounds. Indeed, the smallest larvae (or leptocephali) of the eel drift for thousands of miles in the Atlantic (Schmidt, 1922).

The needs of larval and adolescent fish differ greatly from those of the adults. Immature fish usually live in shallower water and feed on smaller creatures than do the adults. An example of the extreme contrast between the different environment of adults and immatures is found in eel (*Anguilla* sp.) and salmon. Immature eels, after metamorphosis, live in freshwater, and adults spawn in the ocean, whereas immature salmon live in the ocean and the adults spawn in freshwater. Heincke's law (1913), as applied to plaice, provides another and simpler expression of the difference between the environments of immatures and adults: larger plaice live in deeper water, for as the little plaice move away from their nursery grounds on the beaches, they spread into the deeper water and eventually join the adult stocks on the feeding grounds. Beverton and Holt (1957) described this migrationary spread mathematically as a diffusion. Despite the contrast between the environments of juveniles and adults and despite the great distances over which the larvae and juveniles drift from spawning ground to nursery ground, the spawning grounds tend to be fixed.

As will be shown below, the Pacific salmon returns to its native stream to spawn, perhaps even to its native redds. Figure 6 (in Chapter 1) shows the restricted spawning grounds of the Downs autumn spawning herring in the Southern Bight of the North Sea; that near the Sandettié Bank is only 2000 m along the tidal streams and 500 yards across them (Bolster and Bridger, 1957). On Ballantrae Bank in the Firth of Clyde, in southwest Scotland, there is an equally small spawning ground to which spring spawning herring return year after year (Parrish et al., 1959; Stubbs and Lawrie, 1962). Runnstrøm (1934, 1936) distinguished a number of discrete spawning grounds of the Atlanto-Scandian herring off the Norwegian coast. The herring spawn on gravel beds on a number of small grounds within an area, from which the larvae drift in the same direction.

Plaice spawn in the center of the Southern Bight between the Thames and the Rhine (Beverton, 1962); they have spawned at the same position since 1911. On average the females spawn at a distance apart of 80 m, more densely at the center of the distribution and less so at the edges. One effect of this is that a large patch of eggs (and later larvae) is produced which retains its identity on the larval drift from spawning ground to nursery ground. The Arcto-Norwegian cod stock spawns in a narrow band on the edge of the deep water south of the Lofoten Islands in the Vest Fjord in northern Norway (Fig. 25, Chapter 3). Off Iceland, cod spawn in relatively shallow water between the Westman Islands and the Reykjanes Ridge (Schmidt, 1909). In the North Sea, cod spawn on particular grounds — in the Southern Bight, off the Dogger Bank, off Denmark, and off the Scottish coasts (Daan, 1975). In temperate waters fish tend to spawn at the same positions each year. The spawning grounds may be small, like those of the herring, or extensive, like those of cod and plaice. From each there is a larval drift to the nursery ground in a steady current of the same direction. In tropical and subtropical waters, spawning grounds are less well defined, as are larval drifts and nursery grounds; tuna larvae are found all over the subtropical Pacific at nearly all seasons (Matsumoto, 1966), and in upwelling areas the points of upwelling shift with the angle made by the wind to the coast.

Because bigger fish live in deeper water, a nursery ground on the beach or in shallow water is relatively secure from predators. As the little fish grow they move off into deeper water; Beverton and Holt (1957) analyzed the stock densities of immature plaice by age and distance from the coast and estimated this spread away from the coast into deeper water as about 70 miles in three and a half years. Adolescent fish often recruit to adult stocks on the adult feeding ground. For example, the immature arctic cod, a population of the Atlantic cod living in the Barents Sea, have only to move from the top of the Svalbard shelf to join the adult stocks in the deeper water on the slope of the shelf. Likewise, the nursery grounds of the North Sea herring lie near the adult migration routes between spawning ground and feeding ground (Burd and Cushing, 1962).

Adults of both cod and herring migrate to a spawning ground from which the larvae and juveniles drift to the nursery ground, which supplies the adult stock with recruits in later years. Figure 10 shows the migration circuit in diagrammatic form. Because the adults swim from feeding ground to spawning ground, allowing the drifted larvae to recruit to the feeding ground, the range of migration is roughly the range of larval and juvenile drift. Leptocephali drift across the Atlantic from the Sargasso Sea, to which they return as spawning adults. Cod larvae have drifted from Iceland to West Greenland, and adults have migrated back to Iceland (Tåning, 1937). Pacific hake migrate from the coasts of Oregon and Washington, on the west coast of the United States, to the coast of southern California, and their larvae and juveniles must return to the north (Alverson and Larkins, 1969). Plaice larvae and plaice juveniles drift and diffuse northward for relatively short distances, and the adults must return over the same ground to spawn.

To reach their spawning grounds the adults swim against the current that carries the larvae, a contranatant migration. The word "contranatant" means that, to reach the spawning ground, adult fish must apparently move against those currents that drift the larvae away from it. Some form of movement against the current, directing the migration, is not included in the connotation. Not only are there possibly undescribed countercurrents, but migration by an oriented movement against a current requires that the fish refer their movements to external objects independently of the current. Such referents are sometimes hard to find.

The migration circuit also represents the development of the cohort in the life cycle from hatching to the repeated spawnings of adult life. With the spawning ground of temperate species fixed in position, it follows that the larval drift between it and the fixed nursery ground in a steady current is also fixed in a geographical sense. A long larval drift implies a long spawning migration. In general, larvae metamorphose on reaching the nursery ground,

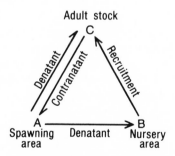

Figure 10. Diagrammatic migration circuit used by fish. Adults apparently migrate against currents from feeding ground to spawning ground, and the larvae drift to the nursery ground. From there the adolescent fish can recruit to the adult stock on the feeding ground. Adapted from Harden Jones, 1968.

although Marliave (1975) has shown that the physiological and structural changes do not match the arrival on the nursery ground exactly. Another important change is the departure from the plankton, from one ecological system to another; perhaps as a consequence the growth rate decreases sharply, in cod from about 10 percent per day to 2 percent per day, in plaice from about 6 percent per day to 1.5 percent per day, and in herring from about 4 percent per day to about 1 percent per day.

Because of the drift of larvae and juveniles and the contranatant movement of the adults, the stock is contained within a constant circuit so long as the spawning ground is always in the same place. The arctic cod larvae drift away from the Vest Fjord in northern Norway, where they were hatched, riding the West Spitsbergen current and the North Cape current (Fig. 11) for 640 km until they reach the Svalbard shelf and the banks of the southeastern Barents Sea, where they settle (Corlett, 1958). The adolescent fish move off these banks into deeper water (Maslov, 1944) and join the adult stocks that live there just before the spawning migration starts, in autumn. Then, contranatantly, the adults travel apparently against the West Spitsbergen current and the North Cape current at 7–8 km/day until they reach the Vest Fjord and spawn (Trout, 1957). It is possible that they ride in a countercurrent on the shelf edge below the Atlantic streams at about 9 km/day (Eggvin, 1964). The Pacific hake may make use of a countercurrent in the Californian upwelling system in order to return from their spawning ground off southern California in spring as the upwelling season progresses (Alverson and Larkins, 1969). The seasonal use of the currents by the fish has been called the hydrographic containment of the stock. Larvae arise on a single spawning ground and, if this ground is always used, a coherence in stock is maintained from generation to generation. Then the stock is a unit population, or, as fisheries biologists say, a unit stock. The characteristics of a stock, its reproductive isolation, and the methods of separating stock from stock will be described in the next chapter. Early fisheries biologists separated Faeroe cod from Iceland cod and from Arcto-Norwegian cod as stocks merely because there were very deep waters between the Faeroes and Iceland and between the Faeroes and Norway, across which the cod would not swim. A cod has, however, crossed the Atlantic (Gulland and Williamson, 1962), but the suppositions of the early fisheries biologists were right: separate stocks are involved, and for the simple reason suggested.

Before the migrations of particular fish species are described, the cruising- and maximum-speed capacities of fish should be noted. Fish cruise at 3 fish lengths per sec, and their fastest speed is about 10 fish lengths per sec (Blaxter and Dickson, 1959; Bainbridge, 1960). Thus, a 70-cm cod cruises at about 4.2 knots and a 25-cm herring at about 1.5 knots. However, Weihs (1973) has suggested on hydrodynamic grounds that the optimal cruising speed might be less, about 1 length/sec. In the southern North Sea, tidal streams move swiftly, up to 2.5 or 3 knots. A cod could cruise more quickly than this during the entire tidal cycle; a herring could cruise more quickly for part of it; a 12-cm sprat (*Clupea sprattus* Linnaeus) could never compete with the tidal streams except at slack water. If, however, such fishes cruise more slowly than 3 lengths/sec they are all at the mercy of the tidal streams in the southern North Sea, and even in the open ocean they might find advantage in riding a current rather than swimming against it.

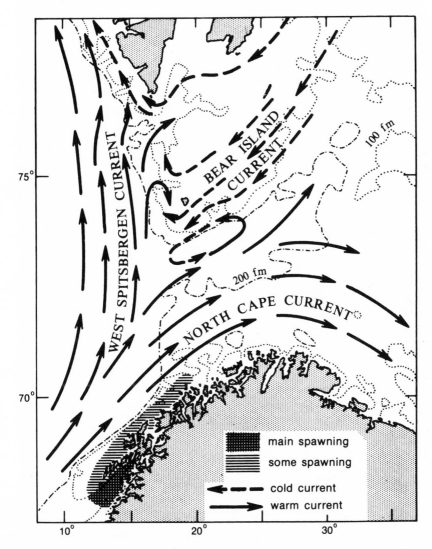

Figure 11. The current system off northern Norway. The arctic cod larvae ride the West Spitsbergen current and the North Cape current until they reach the Svalbard shelf or the south-eastern Barents Sea. Adapted from Corlett, 1958.

The Atlantic Eel

The study of the eel's migration was the first piece of work undertaken by Schmidt (1914, 1915) as the Danish part in international investigations of the North Sea during the first years of this century. (A good summary of eel biology is given in Tesch, 1977.) Schmidt examined the vertebral counts from samples throughout Europe and found that their means were all the same (Schmidt, 1914, 1922). His conclusion was that a single stock supplied eels to all European rivers. Therefore, when the ripening eels go down to the sea, they must spawn out in the ocean, at some distance from the long European coastline. Accordingly, Schmidt made larval surveys outward into the Atlantic and charted the spawning grounds as the places where the smallest leptocephali were caught (Fig. 12), some of which still had yolk-sacs. Unexpectedly, he found that the American eel (*Anguilla rostrata* [Lesueur]) also spawned in 20 m to 500 m in spring and early summer in roughly the same

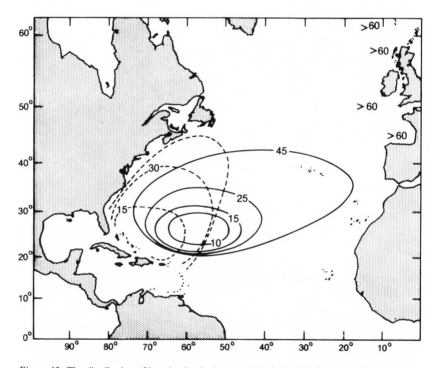

Figure 12. The distribution of larval eels, the leptocephali, in the North Atlantic. The solid line represents the distribution of leptocephali of *Anguilla anguilla*, the dashed line that of the leptocephali of *A. rostrata*. The contours represent areas within which larvae less than 10 mm in length, for example, were found. The patches of smallest larvae occurred in the Sargasso Sea. Adapted from Schmidt, 1922.

area of the Sargasso Sea as the common or European eel (*Anguilla anguilla* Linnaeus). The spawning ground of the American eel lies some distance west of that of the European eel (Fig. 12). The two species are distinguished only by a difference of seven vertebrae in their vertebral counts. Distributions of vertebral counts (Fig. 13) — a mean of 114.7 ± 1.3 for *A. anguilla* and one of 107.2 ± 1.3 for A. *rostrata* — indicate very little overlap between the two

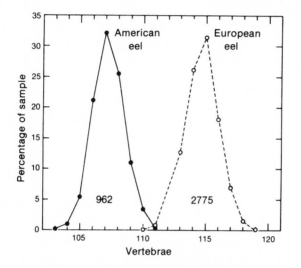

Figure 13. The distributions in vertebral sum for the European eel (*Anguilla anguilla*) and the American eel (*A. rostrata*). Adapted from Harden Jones, 1968.

species (Harden Jones, 1968). This is virtually the same figure as that published by Schmidt (1914), but Ege (1939) had a greater number of observations on the American eel. This difference in number of vertebrae is a large one, much greater than can be generated merely by environmental effects. Schmidt (1917) made extensive studies of the effects of temperature differences on the vertebral counts of *Zoarces*, a genus of the blenny family. From this work he was able to distinguish environmental and genetic effects, and accordingly to imply that the big difference in vertebral count found between the two eel species was a genetic one (see Chapter 3 for a description of the recent work on this problem).

The American eel grows quickly and metamorphoses after a year, whereas the European eel grows slowly and metamorphoses after two and one-half years (Ehrenbaum and Marukawa, 1913; Fig. 14). The young European eel drifts at the surface and appears to eat diatoms and coccolithophorids. Inasmuch as the drift of water across the Atlantic takes about two and one-half years (Iselin, 1936) and the leptocephali seem to remain within this drift, there is a correlation with the probable ages of the young eels; thus, the slow

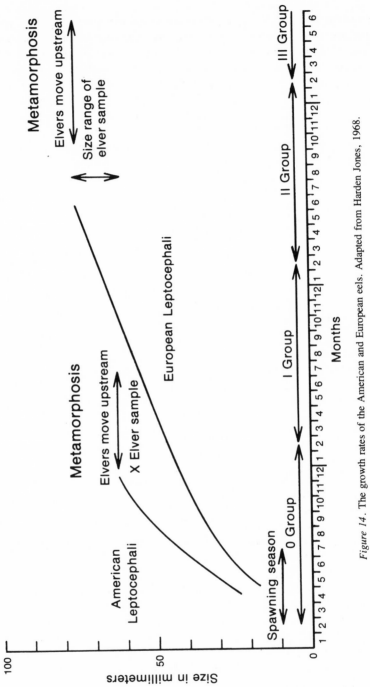

Figure 14. The growth rates of the American and European eels. Adapted from Harden Jones, 1968.

growth to metamorphosis of the European eel is apparently associated with the time needed to cross the ocean. The faster-growing American eel does not have so far to go and the leptocephali metamorphose within the year. It is worth contrasting the growth of the European eel to a few g in three years with that of some tuna species (e.g., Bluefin, *Thunnus thynnus* [Linnaeus]), which in the same time grow to well over 50 kg (Tiews, 1963)). After metamorphosis, the eels move up the rivers, where they may live for many years. Toward the end of this phase of their lives, the adults turn from yellow to silver, their gonads ripen, the gut shrivels (Grassi, 1896), and the retinal pigment turns to deep-sea gold (Carlisle and Denton, 1959). Then the fish are ready to put to sea. Presumably they swim to the Sargasso, to spawn there and die.

The adult migration is, or rather must be, of the same magnitude as the larval drift. The larval drift is slow, and growth of the leptocephali is astonishingly slow during their very specialized existence. The stocks of both eel species are contained oceanographically within the North Atlantic gyre; leptocephali of both the American and European species ride the current systems to the coasts of the two respective continents. It is not known how they segregate, or how the adults return to the Sargasso Sea. They might drift south with the Canaries current and then west with the North Atlantic gyre, or they might move in the slow countercurrent under the Gulf Stream (Swallow and Worthington, 1961).

Schmidt's presumption of a single European stock was probably right, justified by its containment in the North Atlantic gyre. It was not, however, a necessary conclusion from the consistency of the mean vertebral counts of eels from most of the European rivers. The combination of single spawning ground at the root of the North Atlantic circulation system and homogeneous vertebral counts provides good evidence of there being a single stock.

The Pacific Salmon

Fish of the Pacific salmon group spawn on the gravel redds in rivers during the autumn and winter. Their larvae take sixty days to hatch (Gilbert, 1914), and the young sockeye may spend as much as a year or so in lakes and rivers before swimming down to the sea as smolts; pink salmon, however, put to sea within three months of hatching. Smolts of the Atlantic salmon (*Salmo salar* Linnaeus) may move out to sea at two or three years of age, and in some cold Norwegian rivers smolts do not come down to the sea until they are six years old (J. W. Jones, 1959).

The Pacific salmon are fished by the Japanese south of the Aleutian Islands, over a thousand miles from land, east of 180° W. Sockeye (*Oncorhynchus nerka* [Walbaum]), chum (*O. keta* [Walbaum]), pink (*O. gorbuscha* [Walbaum]), Chinook (*O. tshawytscha* [Walbaum]), and coho salmon (*O. kisutch* [Walbaum]) are caught in this area from the open Pacific by drift nets 50–80 km long (Fukuhara, 1955). Most come from Asian rivers, but a small proportion come from Bristol Bay, Alaska, from the rivers of southeast Alaska, British Columbia, Washington, and Oregon, as shown by tagging results (International North Pacific Fisheries Commission, 1966), and by the distribution of internal parasites — one parasite characteristic of Asian rivers and the other of the rivers of North America (Margolis et al., 1966). Most North American fish are caught in the estuaries and the rivers as the fish return to spawn.

The notable feature about salmon migration is that the fish appear to return to spawn in the stream in which they were hatched. This parent-stream hypothesis is supported by a large quantity of evidence, two examples of which are illustrated here — one from work done at Cultus Lake in British Columbia and the other from McClinton Creek in British Columbia. Table 1 summarizes the life history of the sockeye by seasons for the four years of life

Table 1.   The return of salmon to Cultus Lake (Harden Jones, 1968, after Foerster, 1936)

Season	Year			
	1929	1930	1931	1932
Winter	Hatched	In lake	At sea	At sea
Spring	In lake	To sea	At sea	At sea
Summer	In lake	At sea	At sea	At sea
Autumn	In lake	At sea	At sea	Return to spawn

before it spawns and dies. Cultus Lake, where the sockeye spawn and eventually return, drains into the Fraser River. R. E. Foerster directed large-scale marking experiments in which 660,000 smolts were fin-clipped. While only 2.9 percent of the downstream migrants were subsequently recovered in the approaches to the Fraser River or in Cultus Lake itself, no strays were found in any other river or lake (Harden Jones, 1968, after Foerster, 1936); it is this fact that supports the parent-stream hypothesis. Further, 641,000 out of the original number tagged did not return.

Table 2 summarizes some work done on the pink salmon in McClinton Creek (Pritchard, 1938, 1939). The pink salmon spawn much closer to the sea than the sockeye, and their offspring put to sea as fry; after more than eighteen months at sea, the pink salmon return to their parent streams to spawn and die. In 1934, 2,941 marked fish returned to the creek from a tagging there in 1933; 324 were caught in the fishery at Massett Inlet on the way from McClinton Creek. Of 11 marked fish recovered elsewhere, 7 were taken far from the creek — i.e., a stray percentage of 0.2. These constitute a very small proportion indeed, and Pritchard (1948) shows that such straying to other streams is very much less than might be expected from a random search of other rivers by the fish. These facts also support the parent-stream hypothesis. Another conclusion from the data is that the vast majority of salmon are lost at sea, presumably eaten, or perhaps caught.

An interesting experiment was carried out on the degree of stray of steelhead trout (*Salmo gairdneri* Richardson) between two creeks, Waddell and Scott, five miles apart, in California. Smolts were fin-clipped on their downstream migration; 3 percent returned to Waddell Creek and 0.07 percent strayed to Scott, whereas 2.1 percent returned to Scott Creek with a stray of 0.03 percent (Shapovalov, 1937; Taft and Shapovalov,1938). With parent streams so close, the proportion of stray was 14–25 percent, which was much higher than that of the pink salmon from Massett Inlet. Clutter and Whitesel (1956) discovered that they could distinguish spawning groups (Cultus, Fran-çois, Shuswap, and Stuart) within the Fraser River system; the groups differed in the number of circuli (or non-annual rings) on their scales. Henry (1961) identified a larger number of groups in the fishery in the Juan de Fuca Strait (between Vancouver Island and the mainland of Canada and the United States); they pass through the fishery in the same sequence each year, week by week, those from the most distant streams first, and these fish swim most quickly up the Fraser.

Very little of this remarkable mechanism has been revealed, but Harden Jones (1968) has summarized much of the available information. Convinced that fish at sea have no obvious external referent, he believes that the Pacific salmon drifts near the surface of the Alaska gyre till it meets an odor spread from its parent stream and swims toward it; this view has been disputed but remains the simplest explanation. Hasler and his co-workers showed that salmon, among other fish, might respond to olfactory clues during their migration (Hasler, 1966; Cooper et al., 1975). Harden Jones extended this idea in

Table 2.   The return of salmon to McClinton Creek (Pritchard, 1938, 1939)

Movement pattern

Year 1	Autumn	Autumnal spawning run
Year 2	Spring	Spring migration of fry to the sea
	Autumn	No autumnal spawning run
Year 3	Spring	No spring migration of fry to the sea
	Autumn	Autumnal spawning run of spring downstream migrants of year 2

Marking and recapture data

Spring downstream migrants			Autumnal adult return		
Year	No. unmarked	No. marked	Year	No. unmarked	No. marked
1931	5,200,000	185,057	1932	15,504 (0.30%)	96 (0.05%)
1933	2,150,000	107,949	1934	152,255 (7.08%)	2,941 (2.72%)
1935	12,500,000	85,634	1936	52,277 (0.42%)	35 (0.04%)

his sequential hypothesis; the Fraser River system is a large and complicated one, and any returning salmon must face many tributaries before the right sequence is chosen. Harden Jones thinks that olfactory clues are laid down in sequence on the downstream migration to the sea and that when the fish return from the ocean to spawn the sequence is played back until they reach the parent redds.

This brief survey indicates, then, that the adult salmon migration is probably of the same range as the larval and juvenile drift; this hypothesis assumes that the smolts drift down the Fraser River and around the Alaskan gyre. The adult fish segregate to their parent stream, and the stock is that body of fish entering the stream as adults or as smolts leaving it. If the salmon from different rivers live together at sea, the oceanic fishing mortality would be generated randomly with respect to the different rivers; and if this loss were the largest component of fishing mortality, it might be practical to treat the salmon from the Pacific coast of North America as a single stock. But such a concept might become dramatically difficult to apply to separate groups like those from Cultus Lake or McClinton Creek, in which mortality would vary capriciously.

North Sea Plaice

In the southern North Sea, there appear to be three spawning groups of plaice — in the German Bight, in the Southern Bight, and in the Flamborough Head area (Figure 15; Simpson, 1959). This simple figure summarizes a very large quantity of material collected at sea since 1911. In the Southern Bight, the spawning season lasts from December to March, peaking in January; in the Flamborough Head area, it lasts from January to March, peaking in March; in the German Bight, it lasts from January to April, peaking in February. The spawning seasons between the three areas differ by about a month. Harding et al. (1978) showed that since the late fifties and early sixties the stock extended its spawning area considerably to the north and east of the Dogger Bank as it increased in abundance in the German Bight. From the Southern Bight, currents carry the larvae to the continental coast; after five or six weeks, they settle to the bottom and metamorphose, living almost among the breakers (Simpson, 1959).

Figure 16 (Cushing, 1972a) shows the direction of larval drift from that spawning ground between the Thames and the Rhine and the Texel gate, where the metamorphosing larvae near the seabed pass through the sharp boundary between the main stream and the Dutch coastal water. Where the tidal streams flow parallel to this boundary, the bottom water moves shoreward during part of the tidal cycle (Dietrich, 1954). Harding et al. (1978) found that metamorphosing plaice were concentrated in the shallows off Texel Island. Zijlstra (1972) showed that the nursery ground of this spawning group lay on the broad flats of the Waddenzee, within the Frisian Islands. One of the most remarkable points about this larval drift is that it passes into somewhat shallower water where the production cycle must start a little earlier than on the spawning ground itself, and so the larvae obtain the best chance of finding food.

The "postage stamp" plaice live on the beaches between April or May of their first year and the first months of the following year. As juvenile fish, they spread outward from the coast, and in three years reach a depth of 20 m and a size that ranges in length from 10 cm to 23 cm (Heincke, 1913). It is likely that the recruit fish join the Southern Bight spawning group of adults on their way south to spawn or on the spawning ground itself. This is a case where the

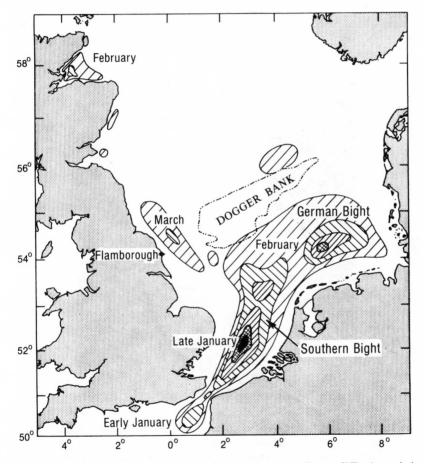

Figure 15. The three spawning areas of plaice in the southern North Sea — off Flamborough, in the German Bight and in the Southern Bight. The contours represent densities of plaice eggs, and the broken line around the Dogger Bank outlines the 20-fm line. Adapted from Simpson, 1959.

fish recruit from the nursery ground to the spawning ground rather than the
feeding ground.

A considerable advance in the methods of migration studies has been made
by Harden Jones and his co-workers with the Admiralty Research Laboratory
(ARL) sector scanning sonar and with transponding acoustic tags. The scanner is a high resolution sonar with an operating frequency of 300 kHz, a range
resolution of 7.5 cm and an angular resolution of 1/3° in one dimension of the
beam and of about 5° in the other. The beam is scanned in angle by steps of
7.5 cm outward in range, and the received signals are presented on an oscilloscope with bearing in one dimension and range in the other. With a range of
about 400 m, very detailed pictures can be obtained of fish shoals, sand waves
on the seabed, and of shipwrecks. Single plaice fitted with small acoustic
transponding tags (Mitson and Storeton West, 1971) have been followed by
sector scanning sonar for as long as 54 h and over distances of up to 61 km.
Plaice moving from one position to another came off the bottom at slack

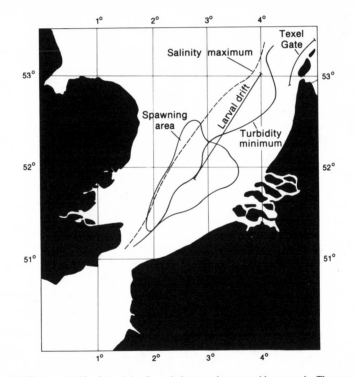

Figure 16. The larval drift of the plaice from their spawning ground between the Thames and the
Rhine toward the Texel Gate, where they enter the nursery ground on the flats of the Waddenzee.
The larval drift lies in the clearer and more saline water. The water is shallower toward the north,
where the production cycle starts earlier, so the larvae are drifted toward food. Adapted from
Cushing, 1972a.

water, moved downstream with the tide in midwater, and returned to the bottom at the next slack (Harden Jones, Greer Walker, and Arnold, 1978). Midwater fishing experiments along the lines of the migration route of plaice in the Southern Bight of the North Sea have shown that selective tidal transport is used by plaice, ripe fish moving south in midwater on the southgoing tides, and spent fish moving north on the northbound tides (Harden Jones et al., 1979). The tidal streams in the southern North Sea are swift, and the plaice appear to have evolved a mechanism by which they can migrate for 200–300 km to the spawning area and back again with the least expenditure of energy; indeed up to 90 percent might be saved (Weihs, 1978).

Plaice of the German Bight and the Southern Bight have been studied by de Veen and Boerema (1959), who found a pronounced difference between the two spawning groups in the length distributions of the nuclei marked on the otoliths (Fig. 17). This difference between the two distributions suggests that there is little mixing between the two groups, particularly as the otolith is laid down at an early stage of the life cycle, during the larval drift. Plaice tagged on the German Bight or Flamborough spawning grounds return in the following year to spawn on the original ground. No interchange takes place among adults between the three spawning grounds, even after a number of years. The same phenomenon, which occurs for the plaice after two years of age, has also been demonstrated for soles (*Solea solea* [Linnaeus]) (de Veen,

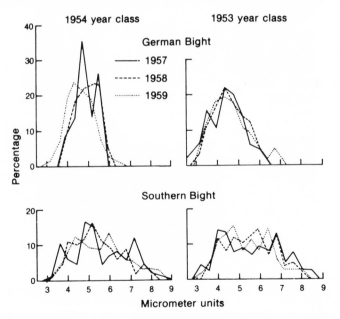

Figure 17. Diameters of otolith nuclei of plaice found spawning in the German Bight and in the Southern Bight. Adapted from de Veen and Boerema, 1959.

1961). The possibility remains that a small degree of mixture between the three groups occurs in the recruitment mechanism.

The adult migration of the plaice is of about the same range as that of the larval drift. Adult fish return to the same single spawning ground year after year and do not move to other spawning grounds. Examination of the diameters of otolith nuclei reveals marked differences, and suggests that the majority of plaice spawn on their native grounds. There is a resemblance to the parent-stream hypothesis, but it is limited to the return of adults to the same spawning ground year after year because the connection between generations cannot yet be established. For the salmon, this connection was made by fin-clipping the young smolts on their way downstream.

North Sea Herring

North Sea herring are predominantly autumn spawners of four spawning groups. Spring spawners and local races exist (Wood, 1937), but they are of small importance. Many complexities arise because, although adults of the four groups probably segregate to their separate spawning grounds, they mix on a common feeding ground. Eggs are laid on grounds of narrow patches of rough sand or gravel inside the 80-m line and, after hatching, the larvae are drifted to coastal areas (Fig. 18), where the whitebait herring (fish of 5–8 cm) live. The four groups are: (1) the Buchan group, spawning from August to September between the Orkney Islands and the Turbot Bank in the northwestern North Sea (Parrish and Craig, 1963); (2) the Dogger Bank group, spawning from September to October around the slopes of the Dogger Bank in the central North Sea (Gilis, 1957); and (3) the Downs group, spawning from November to January in the Straits of Dover and in the English Channel (Cushing and Burd, 1957; Cushing and Bridger, 1966). There is a fourth group, the West Scotland group, which spawns off the Hebrides in August and September.

There are differences in meristic characters between the groups in addition to differences in recruitment, growth, and mortality. A meristic character is one differentiating two populations by the difference between means of a large number of measurements, like the difference in mean vertebral count between the European eel and the American one. The most important difference, however, is in the time of spawning, because, from year to year, the time of spawning on any one ground appears to be very regular and precise.

The whitebait live along the Dutch, German, and Danish coasts in great numbers (Bückmann, 1942), very close inshore. At about one year of age they move out beyond the 20-m line to the Bløden ground (Bertelsen and Popp Madsen, 1953–57), which is an extensive shallow area east of the Dogger Bank. They remain here until they are about two and one-half years old, and during this period are exploited by an industrial fishery. Figure 19 shows how the adolescent fish may recruit to the adult stocks from the Bløden ground. A

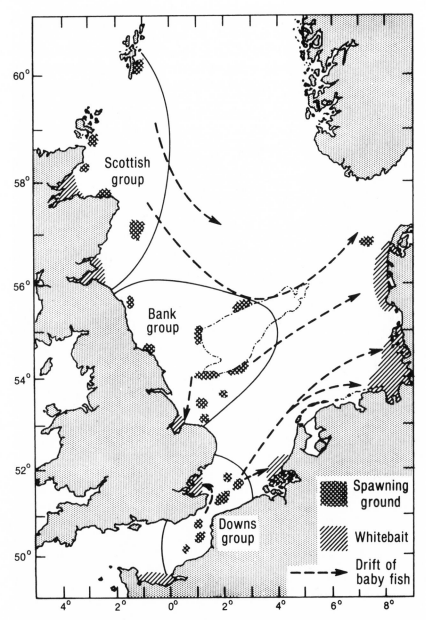

Figure 18. The three groups of herring spawning grounds in the North Sea — the Scottish (or Buchan) group, the Dogger Bank group, and the Downs group. The possible drift of larvae is shown by the dashed lines. Diagonally hatched areas indicate where whitebait are found; the most important are those off the Danish coast.

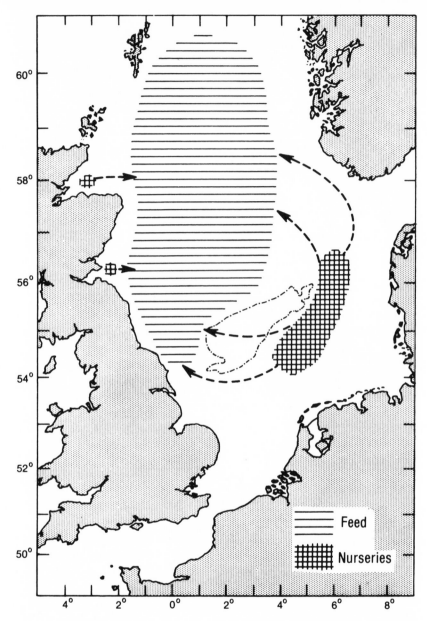

Figure 19. Possible migration routes of adolescent herring from the nursery ground east of the Dogger Bank to the feeding grounds between the Dogger Bank and the Shetland Islands in the north. The Dogger Bank is delineated by a depth contour.

group (which may be Dogger fish) moves north toward the Norwegian deep water in autumn, eventually joining the adult population in the northern North Sea; a younger group (which may be Downs fish) moves around the Dogger Bank in the following spring, eventually joining the adult population in the central North Sea (Burd and Cushing, 1962).

Movements of the adult fish are not exactly known. Höglund (1955) has tagged herring among the skerries on the Bohuslän (southwestern) coast of Sweden in January and February. The skerries are rocky islets extending up to three miles off the coast, and the herring are caught in pound nets from them. Because of their having been taken in these nets, they were in very good condition when tagged. Although Höglund's data are scanty for the winter and spring months, it appears that the timetable of their return was roughly a circuit of the northern and central North Sea (Table 3; Fig. 19). The direction

Table 3. The North Sea circuit of herring as indicated by the recapture of individuals tagged in the Skagerrak (Höglund, 1955)

Direction of recovery from Skagerrak	Month							
	July	Aug.	Sept.	Oct.	Nov.	Dec.	Jan.	Feb.
Northeast and north*	1	13	6					
Central†			23	24	5			
South and east‡				1	3			1

* Egersund Bank (northeast) and Fladen (north).
† Gut, Dogger, Whitby, and Western Hole.
‡ Sandettié (south), Horns Reef, and Hantsholm (east).

of movement is that of the drift of water around the main North Sea swirl. At three miles per day (Tait, 1930, 1937), herring might drift around the North Sea in one year with some time to spare, provided they could change from current to current. For example, Downs herring live in the main North Sea swirl in summer, but in the Channel stream in winter.

Migratory movements in the North Sea involve three adult stock groups (Fig. 20). Those of the Dogger and Buchan groups are roughly based on Höglund's tagging data and the presence of Buchan fish in the Skagerrak. The presence of Downs fish in the northern and central North Sea has been demonstrated through use of meristic characters and age distributions obtained from fish on the Fladen ground in the northern North Sea (Wood, 1937; Krefft, 1954). Burd (Burd and Cushing, 1962) has correlated the stock densities of two-and-one-half-year-old fish at North Shields on the northeast coast of England with those of three-year-olds at East Anglia. Similarly, he has correlated three- at East Anglia with three-and-one-half-year-old fish at Shields in the following summer, and three-and-one-half-year-old fish at Shields with four-year-olds at East Anglia in the autumn. Hence, it would seem possible that the Downs fish move north and south off the English coast (Burd and

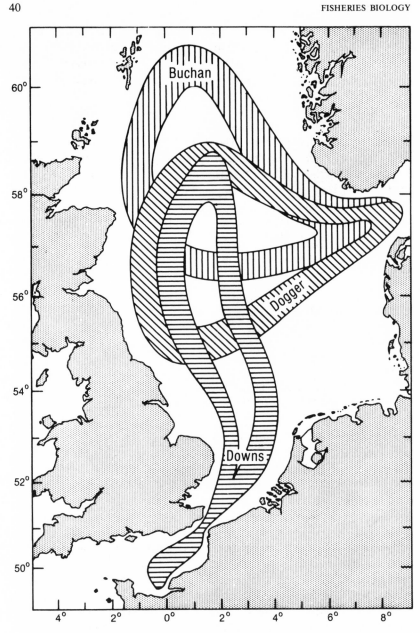

Figure 20. Possible adult migration routes of the three spawning groups of herring in the North Sea. The most northerly one is the Scottish (Buchan) group, the middle one the Dogger Bank group, and the southerly one is the Downs group. Adapted from Cushing and Bridger, 1966.

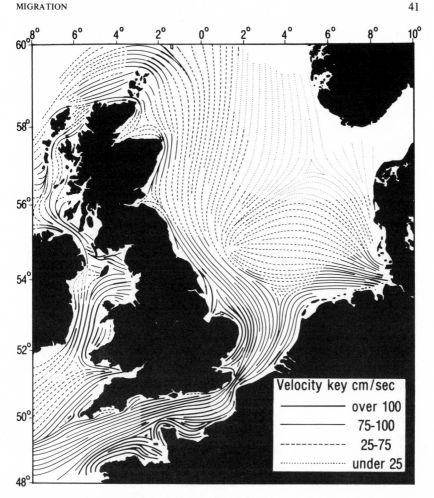

Figure 21. Chart of the direction of tidal streams at one state of the tide in the waters around the British Isles. The velocity is represented by the thickness of the stream lines. Each stream line represents the distance traveled in the tidal cycle and links it to the next one in the same direction. Adapted from Harden Jones, Greer Walker, and Arnold, 1978.

Cushing, 1962). The course of the fisheries in the Southern Bight, described in Chapter 1, would support this view.

Recently, Harden Jones (Harden Jones, Greer Walker, and Arnold, 1978) made a chart of tidal flow in the North Sea. The usual chart shows the direction of the streams at one state of the tide, but that of tidal flow represents the distance traveled in a tidal cycle and links it to the next one in the same direction, the streams of interest to fishes in a tide-dominated area. This chart is shown in Figure 21, and it is immediately obvious that it resembles Figure

20 very closely. If the adults of the three herring stocks migrate in the tidal streams by selective tidal transport, the spawning grounds shown in Figure 18 will segregate the stocks throughout the life cycle. The essential step is that the larval drift of the Downs stock is separated from that of the other two; the Downs whitebait probably live on the beaches of Holland and Germany, whereas those of the Dogger may be drifted to Danish beaches. Then the distinct migrations of the young fish from the nursery grounds in Figure 19 are explained.

Adult herring migrate a comparatively short distance, roughly the same as that of the larval drift. The spawning grounds are fixed in space and time; this fixity is precise enough to suggest the possibility of homing without having to provide further evidence. Zijlstra (1958) has suggested that the consistent difference in vertebral count of herring between spawning grounds over many years means that most of the fish return to the ground on which they were spawned. However, homing cannot yet be demonstrated. But it can be shown (see Chapter 3) that adults of the Downs substock appear to return year after year to the same spawning ground, once they have recruited to it (Burd and Cushing, 1962).

The Study of Migration by Tagging Experiments

Tagged fish are liberated at one position and are recovered at another; during the time between liberation and recapture the distance is the shortest possible migration. Otsu (1960) showed that albacore released off California were recovered off Japan up to a year and a half later, a swift movement across a considerable distance. The Norwegian herring migrate from the spawning grounds off Norway to their feeding grounds off Iceland and at the polar front between Iceland and Jan Mayen (Aasen and Fridriksson, 1952; Devold, 1963). Thompson and Herrington (1930) plotted the number of tags recovered on distance from liberation from their spawning grounds for the Pacific halibut *Hippoglossus stenolepis* [Schmidt]). The spread of such distributions, if collected for long enough, indicates the average range of migration from the spawning ground, irrespective of distance (Fig. 22). An interesting point is that the group which spawns off British Columbia spreads far less than that off Alaska, many hundreds of miles away. Such results are spectacular, but they require the presence of fishermen at the four corners of ocean to collect the tags.

A more detailed analysis of the results of tagging experiments is due to R. Jones (1959). With the use of an experiment on North Sea haddock (*Melanogrammus aeglefinus* [Linnaeus]), he estimated the directional component of velocity, v', from a coastline and the dispersion coefficient a_1:

$$v' = \Sigma r'_n \cos \theta / \Sigma t, \tag{2}$$

where r'_n is the shortest distance traveled,

θ is the bearing from the axis of the coastline,

t is the number of days at liberty;

$$a^2_1 = (1/n) \; [\Sigma(r'_n{}^2/t) - (\Sigma r'_n \cos \theta)^2 / \Sigma t], \qquad (3)$$

where n is the number of fish recovered.

The haddock migrated at 0.076 to 0.82 miles/day and they dispersed at 2.2 to 118 miles^2/day. Saila (1961) used the same method to analyze an experiment on the winter flounder (*Pseudopleuronectes americanus* [Walbaum]) off Rhode Island; they migrated at 0.03 miles/day and dispersed at 0.67–1.10 miles^2/day, but there was little or no directional component to the migration. With random search, 75 percent of the population would reach the coast after

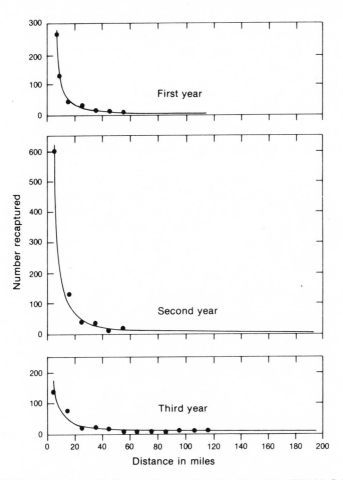

Figure 22. The spread of tagged halibut in miles from the point of liberation off British Columbia for three successive seasons. Adapted from Thompson and Herrington, 1930.

90 days. Errors of the estimates are derived by first transforming polar coordinates to Cartesian ones:

$$x = (1/n) \sum_{i=1} \cos \theta, \; y = (1/n) \sum \sin \theta,$$
$$r'_n = (x^2 + y^2)^{1/2}, \text{ and } \cos \alpha^* = x/r',$$

where α^* is the mean angle at which tags are recovered.

$$s = [2 \, (1\text{-}r'_n)]^{1/2} \tag{4}$$

where s is the mean angular deviation in radians (this statistical treatment is taken from Batschelet, 1965). Any tagging experiment can be used to estimate the directional component of velocity, the mean angular deviation, and the dispersion coefficient provided that the tags have spread in sufficient time across an area adequately sampled by fishing vessels. Such information can be used in models of migration based on random search with small directional components (Saila and Shappy, 1963).

The Nature of Migration

In temperate waters, fish spawn at the same position at the same season each year, and spawning fisheries are equally regular. The spawning season lasts for as long as three months. Usually the older fish with larger eggs spawn first. The peak date of spawning does not vary much (Cushing, 1969a); the average peak date of plaice spawning in the Southern Bight of the North Sea since 1911 has been January 19, with a standard deviation of about a week. Detailed records of spawning have been published by the International Pacific Salmon Commission for about fifty spawning groups or stocklets for seventeen years; they show that the standard deviation of the peak date of spawning is less than a week. The Atlanto-Scandian herring spawns on gravel banks in 40 fms or so off Utsira, in southwestern Norway. Each ground is as restricted as those of the Downs herring shown in Figure 18. They were characterized by Runnstrøm (1936) as "late February" or "early March" grounds; from grab samples for a short period of nine years the standard error of the mean date of spawning was about a week. Between 1890 and 1970 the peak date of spawning of the Arcto-Norwegian cod delayed by about ten days. As climatic amelioration has given way to deterioration, the date has advanced again; however, the standard deviation about the trend was as low as a week. In temperate seas fishes of the large stocks of commercial value appear to spawn at a fixed season.

In tropical and subtropical waters there is no evidence of such precise spawning seasons. Indeed, the standard deviation of the spawning date of the California sardine (*Sardinops caerulea* [Girard]) estimated from a decade of egg surveys (Ahlstrom, 1966) is about seven weeks. Further, tuna larvae are caught all over the subtropical Pacific as if spawning were continuous in space and time, although they tend in fact to be more abundant off the Philippines for nine months of the year (Matsumoto, 1966). Within 50° of either pole, production cycles in the sea are discontinuous because no algae grow in the winter (Cushing, 1976a) and food is most abundant during the spring outburst (autumn outbursts occur in certain places). Within 40° of the Equator, there may be perhaps less food, except in the upwelling areas, but it is produced all the year round. At the boundary between temperate and subtropical regions — as, for example, southwest of the British Isles or on the eastern coast of South America — the distinction may become blurred.

The production cycle in temperate waters varies in time of onset (and in time of peak production) in quantity produced and in the duration of production (Colebrook, 1965); from the observations of greenness made with the plankton recorder the spring outburst appears to vary in the time of peak production by two or three weeks, which is a little less than that recorded by the very few observations with other nets. If fish spawn at a fixed season each year, their larvae will feed well or poorly according to whether the production cycle is early or late. The match or mismatch of larval production to that of their food (Cushing, 1974; Cushing, 1976a; Cushing and Dickson, 1976) may be at the root of the high variability of year-class strength in fishes (see Chapter 9). But the more important point in the study of migration is that fish spawn at places whence currents drift their larvae into productive regions, as does the plaice in the southern North Sea. Another example is provided by the autumn spawning herring of the North Sea, for their larvae are drifted eastward into the only areas of high autumn production in the northeast Atlantic, except for southern Iceland (Cushing, 1967a). The southerly migration of the Pacific hake in spring from the coasts of Washington and Oregon to that off southern California (Alverson and Larkins, 1969) may allow their offspring to return north with the upwelling season, which starts off southern California in February and reaches Oregon in July and August (Cushing, 1971a). Figure 23 expresses such ideas a little more generally in the region of cyclonic circulation within 50° of either pole and in that of anticyclonic circulation within 40° of the Equator. Each diagram represents a migration circuit with respect to the period of production; spawning ground and feeding grounds are indicated together with larval drift and adult migration. In the subtropics, with a continuous production cycle and food always available, the circuits of migration are perhaps a little indistinct, whereas in temperate waters they are precise and pronounced because most food is produced in the somewhat short spring outburst.

Fishes grow considerably as they pass through their life cycle, eat successively larger particles, and rise in the trophic hierarchy of the ecosystem. The growth of adult fishes does not appear to be limited by food lack, but that of the juveniles is density dependent (see Chapter 4); hence the position of the adult feeding ground is unimportant as compared with that of the larval drift between the fixed spawning ground and the fixed nursery ground. The feeding ground may lie at the end of the slow spread into deep water from the nursery ground. During their early period of the life history the magnitude of the cohort is established, and the density-dependent processes may well have been completed. This principle probably holds true even in subtropical waters, if in an imprecise way. Fish migrate across considerable distances in order to feed the adults freely and to establish the position of the all-important larval drift at the right time. Perhaps fish migrate in temperate waters to remain in the same region and to generate the year classes at a fixed and productive position against the pervasive flow of the waters.

   If a fish stock remains in the same region of the sea and is large enough, it can maintain its reproductive isolation. The plaice population in the southern North Sea and eastern English Channel comprises three or four spawning groups with three or four distinct nursery grounds (Harding, Nichols, and Tungate, 1978; Zijlstra, 1972) between which there is potential exchange from generation to generation. A quite distinct stock spawns in the Moray Firth in northeast Scotland, and some of the larvae must drift southwards in the direction of the southern North Sea; with normal drift, however, they would settle to the seabed long before reaching the southern population, and so the two stocks remain distinct. Within the southern stock, most of the

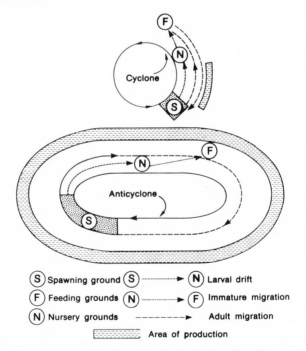

Figure 23. The relation between migration and production in the cyclone of temperate waters and in the anticyclone of subtropical seas. Production in temperate waters stops in winter, but that in subtropical seas continues all the year round. Spawners in temperate waters migrate contranatantly to their spawning grounds, but those in subtropical seas drift continuously, spawning and feeding at the same time. Adapted from Cushing, 1975b.

larvae from the Southern Bight spawning ground reach the Texel Gate (Fig. 16), but some may drift further and reach the coast of Denmark (Ramster, Wyatt, and Houghton, 1975). There may be, however, relatively little interchange between spawning groups within the stock, which also follows from the limited range of the larval drift. We shall return to this concept in the next chapter, on the nature of the unit stock; all we need note here is that a stock is isolated from others that lie beyond the range of its larval drift.

This view of migration in temperate waters, where a stock remains in the same region, also requires that the circuit of migration is contained within a specified current system. In the loosest sense, albacore drift or swim round the North Pacific anticyclone and do not migrate elsewhere (Rothschild and Yong, 1970). The Pacific hake make use of the current systems within an upwelling area to migrate south to spawn off the Pacific coast of the United States. The salmon return from their oceanic journeys south of the Aleutians and leave the Alaska gyre. They may use a sequence of signals to find their way home in the Fraser River system. The Arcto-Norwegian cod use the Atlantic stream or its countercurrent to migrate from the Vest Fjord to the Barents Sea, yet on their spawning migration they can turn from the current system round the island of Röst and enter the Vest Fjord. The hydrographic containment of the fish stock requires that fish of particular ages make use of the current system to maintain the migration circuit for the purposes indicated above; the fish leave or join a current at a place or season and their journey completes a stage in the circuit of migration. How such choices are made in physiological terms remains unknown.

Spawning seasons last for a long time — two, three, or four months — which might minimize the chance of mismatch of larval production to that of larval food. Consequently, the larval drift persists for months and the patches are broad enough to retain their identity against the diffusive processes of the sea; the nursery season therefore also lasts for a long time. Thus, spawning and nursery ground are pre-empted by the single population for long periods, which prevents the emergence of any competing stock. It follows that a population that sustains such a migratory circuit must be a large one. The large stocks that support commercial fisheries have long been observed to be migratory merely in that fisheries occur regularly at the same place and at the same time from year to year.

Summary

Combining the material for four fish species — eel, salmon, plaice, and herring — we may conclude that the adult migration is of the same range as the drift of larvae and juveniles, at least partially across the Pacific, across the Atlantic, or across the North Sea. The spawning grounds appear to be often fixed in space and constant in time. Salmon return to their native streams and eels must return to the Sargasso. Plaice and herring spawn on fixed grounds at the same season each year, but whether they are their native grounds is not known.

In the four fish species examined, water movements drift the larvae to the nursery ground, from which they eventually find their way back to the spawning ground as adults. Because the spawning ground remains in a fixed position, the circuit of migration provides that coherence between generations which is the proper basis of stock analysis.

Migration is described well in the distribution of the recoveries of tagged fish in space and time from the point of liberation. Acoustically transponding tags have been used to describe the movement of individual fishes by selective tidal transport in shallow seas. Tagging has revealed the capacity of fishes to travel considerable distances with little expenditure of energy.

Migration of marine fishes is linked to the timing and position of the production cycle; larvae are released in regions and seasons where food is abundant. Hence the migration circuit must be constructed from the distribution of tides and currents to bring the spawners at the right time and right place to release the larvae into the production cycle.

3

The Idea of a Unit Stock

In temperate and in arctic waters, a fish stock often has a fixed spawning ground, a single short spawning season, and probably a consistent migratory circuit. An ideal unit of population is one in which the chances of mating are randomly distributed. Because fishes gather from great distances to spawn in restricted areas, their unit stocks may be such ideal units. Ideally, spawners do not leave the stock or join it from other spawning grounds to any great extent from year to year; we cannot establish this point directly, but the genetic evidence to be given below suggests that it is true. A species can comprise a single stock (on rare occasions, as in Atlantic eel), but more frequently a species or subspecies comprises a number of stocks, as will be shown below. So a stock is often a unit of population of lower category than that recognized by taxonomists.

In simple cases, e.g., the arctic cod, many of these criteria are fulfilled; indeed, as will be shown below, this stock can be distinguished genetically from a coastal cod stock which shares the same spawning ground. Difficulty in recognizing what comprises a single stock arises where stocks mix at some stage in their migratory circuits, as in the North Sea herring. If stocks could be separated in the mixed populations, it would be possible to classify each fish as belonging to one stock or another. However, taxonomists do not leave such distinguishing characters unnoticed. The distinctions found between fish stocks are sometimes at a lower level of difference than those found by taxonomists between individuals of different species. Attempts have often been made to establish such small differences through the use of meristic characters or the distributions of characteristics such as numbers of vertebrae or keeled scales. This technique often fails because there may be a large overlap in the distributions of two populations which is caused not by mixture but merely by insufficient degree of difference. Statistical methods, such as discriminatory analysis and distance functions (Rao, 1952), can be used to

49

combine the differences in meristic characters found between the spawning stocks of sea fish, but they are difficult to use successfully for the same reasons. However, such methods have been supplanted by the use of allozymes — i.e., proteins which are coded by genes in individuals and which are expressed as the presence or absence of alternative forms of genes. Further, the distributions of different characters can be combined to estimate genetic distance between populations. Again, the combination of estimates with different genetic measures is much easier and yields more information about breeding groups than do meristic characters.

One meristic difference is very clear — the difference in vertebral count between the European and American eel, which is large enough to be clearly of genetic origin, as noted in Chapter 2. The spawning grounds of the two species are relatively close to each other, but they also are distinct, the adults presumably returning to their native grounds (Schmidt, 1922). The migratory circuit contained by the North Atlantic gyre provides the coherence between generations. If spawners do not leave the stock or others join it for a long period, the stock is isolated. The difference between American and European eels is a genetic one, indicating sufficiently long isolation to generate a specific difference (Sick et al. [1967] and Fine et al. [1967] showed differences in transferrins between the two species; Drilhon and Fine [1969] showed that one allele of hemoglobin was present in *Anguilla rostrata* only; Rodino and Comparini [1979], showed much variability in Atlantic and Mediterranean eels, but their evidence supported the homogeneity of the stock.)

In many regions stocks mix on their feeding grounds. Because fish cannot be tagged as eggs or larvae, it is hard to show that they return to their native grounds to spawn. A more limited proposition is that the adult fish return to the same fixed spawning grounds year after year. Then a stock may be defined as a population in which the vital parameters of recruitment, growth, and mortality are homogeneous. Because the population should be examined, not in part, but as fully as possible, it is convenient to make the tests of homogeneity on the spawning populations. Then, between spawning groups, a heterogeneity in the vital parameters would be evidence of true difference. A stock may comprise a number of spawning groups between which interchange is low and within each of which chances of mating are randomly distributed, as noted below. Then the spawning group or stocklet is the natural unit of study. As examples, the Arcto-Norwegian stock of cod, which mixes with no other stock in the Barents Sea, will be contrasted with the Downs stock of herring, which mixes on its feeding grounds in the northern and central North Sea with two other stocks.

The Arctic Cod

By means of differences in two meristic characters, vertebral count and fin-ray count, Schmidt (1909, 1930) separated the Arcto-Norwegian cod stocks from the cod stocks at Iceland and at the Faeroes. The differences were low enough to have been generated environmentally. But a small difference of environmental origin, which persists from generation to generation, must show that most of the fish were spawned separately in the two distinct environments.

The southward migration from Malangen to the Vest Fjord (Fig. 24) has been charted in some detail by echo survey (Saetersdal and Hylen, 1959). With this method, a research vessel can be used to survey an area with its echo sounder. Ultrasonic pulses are transmitted to the seabed once a second, and the machine makes a continuous record of all echoes received between surface and bottom along the ship's track. Echoes from fish on the bottom or in any part of the water column appear on the paper record, and they can be marked on a survey chart as the number of fish traces per mile along the ship's track. Echo surveys carried out by Norwegians on research ships off the northern coast of Norway (Fig. 24, a) show echo traces of large skrei or mature cod from the gullies between the banks, where the British trawlermen had been catching them. The echo traces from cruise to cruise indicate a southerly

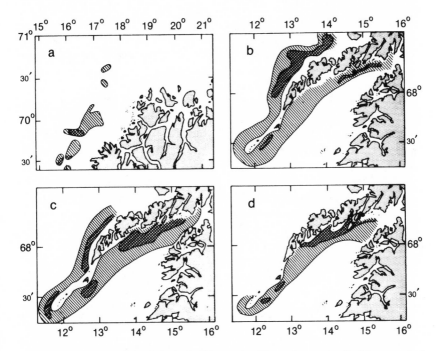

Figure 24. Echo surveys of cod obtained off the northern Norwegian coast as the fish approached the Vest Fjord. The cross-hatched areas are those of highest density; diagonally hatched ones are areas of lower density. Periods of the surveys were (*a*) 29 January–7 February 1959; (*b*) 8–11 February 1959; (*c*) 25–28 February and 2–3 March 1959; (*d*) 13–15 March 1959. Adapted from Saetersdal and Hylen, 1959.

movement of skrei from Malangen around the southernmost of the Lofoten Islands into the Vest Fjord (Fig. 24, *b–d*), where the ripe cod spawn. Echo surveys also show (Fig. 25; Bostrøm, 1955) a narrow patch of fish traces on the north side of the Vest Fjord off the Lofoten Islands. These fish traces lie in a transition layer at a depth of about 70 m on the northern edge of the fjord between the warmer, saltier Atlantic water below and the cooler, fresher runoff from the land. Norwegians call this the "fish-carrying" layer (Rollef-

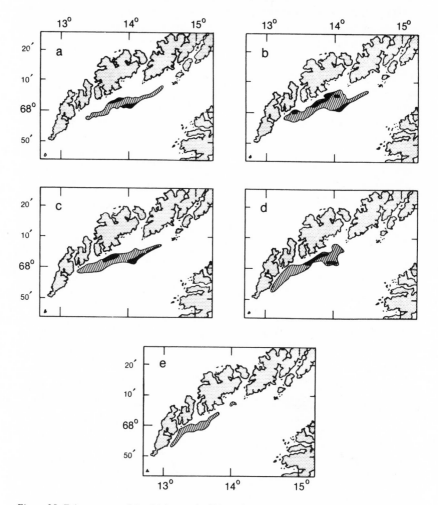

Figure 25. Echo surveys of the "fish-carrying" layer in the Vest Fjord: (*a*) 1–5 March; (*b*) 7–12 March; (*c*) 14–19 March; (*d*) 21–26 March; (*e*) 28 March–2 April. Black areas are those of highest cod density; diagonally hatched ones are areas of lower density. Adapted from Bostrøm, 1955.

sen, 1955). The male fish arrive in the Vest Fjord first, and spawning takes place in the midwater layer where the fish are caught by drift nets, longlines, and purse seines (Rollefsen, 1954) used by fishermen from the port of Svolvaer. By the end of March, spawning has finished and echo surveys indicate the northward movement of the fish. They appear to move out around the southernmost island of Röst in the same way as they entered the fjord, but tagging data suggest that some fish move north between the islands (Hjort, 1914).

The cod larvae are dispersed by the West Spitsbergen current to the Svalbard shelf and by the North Cape current onto the banks of the southeastern Barents Sea (Corlett, 1958). Here the young fish stay and grow into adolescents. These nursery grounds are very close to the adult feeding grounds and the maturing young recruits merely swim into deeper water as they grow bigger and join the adult stocks there (Maslov, 1944) before the southbound spawning migration.

The range of migration by the arctic cod is demonstrated to some extent by tagging. Hjort's experiment (1914) (Fig. 26) shows the northerly movement away from the Lofoten Islands (at 18 miles per day); later recaptures were taken around the northern coast of Norway. This experiment was conducted in 1911, before trawlers visited the Svalbard shelf regularly, and if fish reached

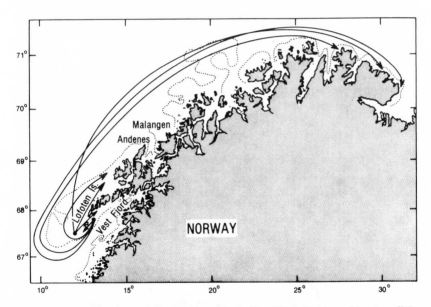

Figure 26. Hjort's tagging experiment with cod in the Vest Fjord. Positions of tagging off the Lofoten Islands in the Vest Fjord are shown as solid circles. The solid lines indicate the direction of movement of the tagged fish to their positions of recapture. The dotted line is the 100-fm contour. Adapted from Hjort, 1914.

there none were recaptured. Dannevig's experiment (1954) (Fig. 27), carried out in 1949, shows a much greater range in positions of recapture from an area southeast of Spitsbergen to the southeastern Barents Sea. The range of recaptures illustrated in Dannevig's chart is limited by biological barriers which the fish do not cross (see Figs. 28 and 29). It is the present known range of the arctic cod in the Barents Sea, and the contrast between Hjort's and Dannevig's data shows the danger of relying too heavily on the distribution of fishing fleets to chart the distribution of fish stocks. In another experiment, Maslov (1944) found that cod he tagged in the southeastern Barents Sea and off Novaya Zemlya in summer were recaptured some 300–400 miles westward by early autumn. Results from still later research, conducted by Trout (1957), demonstrated that cod tagged from the Svalbard shelf off Spitsbergen and Hope Island, lying to the southeast, returned to the Vest Fjord area. Some fish were caught in spawning condition in the coastal waters south of the Vest Fjord, but these comprised only a small proportion. The general conclusion from the four liberations of tagged cod by Norwegian, Russian, and British scientists is that there is a single spawning area, which is predominantly in the Vest Fjord, and that the cod return to this area year after year from their feeding grounds on the Svalbard shelf and in the southeastern Barents Sea. No marked fish liberated on the Svalbard shelf or in the southeastern Barents Sea have been recovered on the spawning grounds off Iceland and the Faeroes where the fish of the nearest stocks, separate from the arctic cod, spawn. The

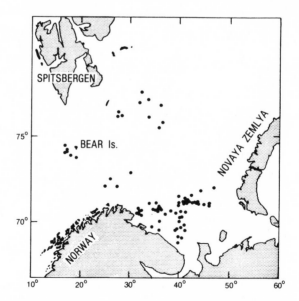

Figure 27. Dannevig's tagging experiment with cod. The fish were tagged in the Vest Fjord, and the solid circles indicate the positions of recapture. Adapted from Dannevig, 1954.

evidence that the stock is a single unit rests on the results of the tagging experiments, on the single discrete spawning ground in February and March, and on the supporting evidence of the echo surveys. The patch of spawning fish shown in Figure 25 is only about 25 miles long by about 5 miles wide and it is likely that most of the spawning population has gathered there from Spitsbergen, the Svalbard shelf, and the southeastern Barents Sea.

The distribution of cod in the Barents Sea has been studied by echo survey. Figure 28 shows three such echo surveys that were made within a fortnight of each other to record bottom-living fish (Cushing, in Richardson et al., 1959). With frequent trawl hauls, the echo patches were identified as cod. The echoes were counted between 0 and 2 fm of the bottom, the depth range fished by the trawl. Signals were not recorded on paper, but on a cathode ray oscilloscope, and the fish echoes were counted and classified by amplitude in μv of received signal. The pattern that emerged as a result of the successive surveys indicated that the cod were apparently "thrown" onto the Svalbard shelf by the West Spitsbergen current rather quickly, and results of a subsequent survey showed

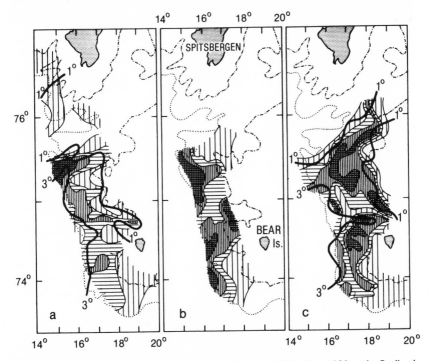

Figure 28. Three successive echo surveys made within a fortnight in June 1956 on the Svalbard shelf. The isotherms of the bottom water moved up the shelf, and with it the echo patches of cod. The four forms of contoured hatching indicate four levels of density, the highest density being shown as cross-hatching. Depth contours are represented by dot-dash lines for 100 fm and dotted lines for 200 fm. Adapted from Cushing, in Richardson et al., 1959.

that the fish had diffused over the whole area of the shelf, away from the West Spitsbergen current at the edge of the shelf, toward Hope Island, about 150 miles to the eastward. The spreading of the current is indicated in the progress of the 2° C isotherm on the bottom up and over the shelf. The important point to notice is that in summer the cod distribution is bounded in all four surveys by the 2° C isotherm on the bottom.

Another isothermal boundary is shown in Figure 29. Four echo surveys, two in spring and two in autumn, were made by Hylen, Midttun, and Saetersdal (1961) off the northern Norwegian and Russian coasts. I have adapted the results from two of these surveys (one in spring and one in autumn) in Figure 29 (*top*) and (*bottom*). It is possible to estimate the size of a single fish from the signal strength of its echo, as recorded by the echo sounder. By means of a paper recorder, the number of echoes from large single fish were counted and charted together in relation to the positions of the isotherms at 150 m. The fish were identified as cod only by their signal strength on the echo sounder, no other fish so large being found in the area. Charts of temperature distribution indicate that, for cod, the opening to the southeastern Barents Sea may sometimes be narrow. Yet results obtained from Dannevig's and Maslov's tagging experiments show that fish must pass through such an opening both in spring on their way eastward and in autumn on their return. The 2° C isotherm in spring, summer, and early autumn describes the edge of the range of the arctic cod off Spitsbergen, on the Svalbard shelf, and in the southeastern Barents Sea.

The evidence that there is a single, or unit, stock of arctic cod is as follows. From the spawning ground, fish migrate to the northern limits of their range to feed, and they return a year later to spawn. There appear to be two hydrographic mechanisms containing the stock in the region where it lives: (1) the fish are carried by the West Spitsbergen current onto the Svalbard shelf, and (2) the southeastern Barents Sea becomes relatively more open to them in early summer as they drift eastward in the North Cape current. The same currents carry the larvae to the nursery grounds, thus ensuring recruitment to the adult stock. Evidence of homogeneous recruitment to the stock in the area has been provided by Lundbeck (1954), who found a high degree of correlation among the year classes at Bear Island, off the coast of the county of Finnmark in northern Norway, and in the Vest Fjord. There are certain differences in otolith type from cod found on the Svalbard shelf (Trout, 1957), but fish with both types certainly return to spawn in the Vest Fjord. There is no evidence of emigration by Lofoten spawners, or any immigration from other stocks. As there is a single spawning ground, shared with the coastal cod, there must be coherence from generation to generation; and the coherence is preserved because the stock is contained within the system of currents. This very simple sum of evidence is the real basis for the stock unity of the arctic cod, and it makes sense of the small differences in vertebral sum and fin-ray count observed by Schmidt (1930) when he compared samples of the arctic cod with those from Iceland and the Faeroes.

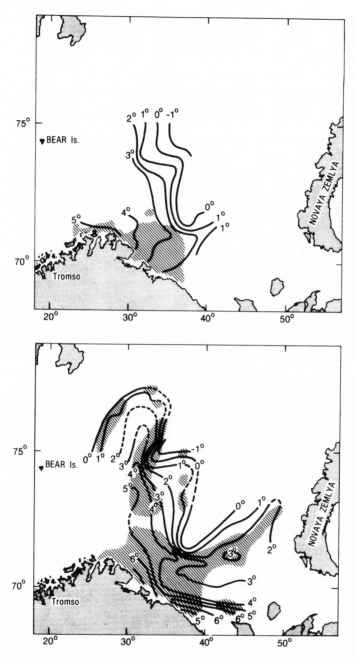

Figure 29. Distribution of cod (diagonal hatching) in the spring of 1960 (*top*) and in the autumn of 1959 (*bottom*) north of the Norwegian coast, as shown from the results of echo surveys. The echoes were from single fish deep in midwater where their distribution is bounded by the isotherms at 150 m. The dashed isotherms are interpolated ones. Adapted from Hylen, Midttun, and Saetersdal, 1961.

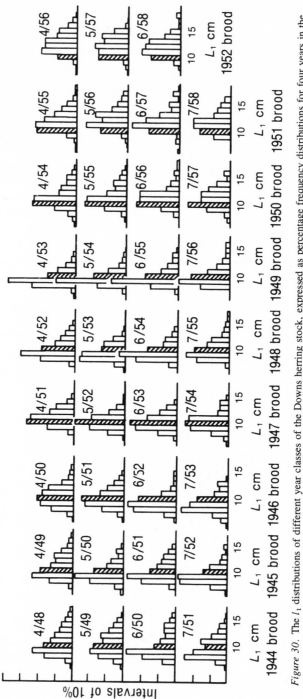

Figure 30. The l_1 distributions of different year classes of the Downs herring stock, expressed as percentage frequency distributions for four years in the lives of the year classes. The 10-cm group is used as a marker to assist comparison between year classes. There is little variation within a year class from four years of age to seven years of age, but there is considerable variation between year classes. Adapted from Burd and Cushing, 1962.

The North Sea Herring

As noted in Chapter 2, there are three autumn spawning groups (and a fourth, if we include the West Scotland spawning group which migrates into the North Sea) of herring in the North Sea — Buchan, Dogger, and Downs — separated by time and place of spawning. There are also differences in mortality, recruitment, and growth among the three groups (Cushing and Bridger, 1966). In addition, there are differences in vertebral count (Cushing and Bridger, 1966; Parrish and Craig, 1963), in body dimensions (Muzinic and Parrish, 1960), and in fecundity (Baxter, 1959). No genetic differences have yet been established between the three groups of spawners (but see Chapter 10); a difference has, however, been found between herring that spawn in the southern North Sea and those that spawn in the eastern English Channel (de Ligny, 1969) at a lower level of population grouping than that between the three major spawning groups. It is possible that the genetic study of herring populations in the North Sea will be fruitful in the future (see Chapter 10).

Because the adult fish of all three groups mix on the feeding grounds in the northern and central North Sea and segregate to their spawning grounds, it must be shown that adult fish of a given year class return to the ground of first spawning year after year. Tagging of adult fish on their spawning ground would normally represent the simplest method of showing the segregation of spawners from feeding ground to spawning ground, but at the present time the loss of tags from North Sea herring is very high, and not enough survive from year to year to demonstrate the segregation.

The problem can be tackled indirectly by using l_1 measurements from the scales. By magnification of the herring scale with a projector, the distance from the nucleus to the first winter ring can be measured. The length at capture is known, and so is the length of the scale at capture. From the ratio of scale length to fish length, the length at the age of the first winter ring can be determined. This measurement is the l_1, the length at which the first winter ring is laid down.

Figure 30 shows the l_1 distributions of year classes within the Downs stock in the East Anglian herring fishery that used to take place in the southern North Sea. The distributions of l_1's are shown for the age groups four to seven in eight year classes (1944–51) and for four to six in the 1952 year class. The distributions within a year class are all very similar, but between year classes there are considerable differences, during a period of enhanced growth; for example, contrast the distribution of the broods 1949, 1950, and 1951.

The mean l_1's of the Buchan and Dogger stocks were 4 cm and 2 cm greater respectively than that of the Downs stock. If they immigrated to the Downs spawning grounds in any significant quantity, the mean l_1 there should increase with age, which it does not. Similarly, there is no evidence from the l_1 distributions of the Buchan and Dogger stocks that there is a decline in mean

l_1 with age as if there had been immigration of smaller fish from the south. The conclusion is that after first spawning, the adult fish return to the same spawning ground year after year. It should be made clear that we are discussing gross changes that might alter the distributions shown in Figure 30 and not the small ones which might destroy any isolation of stocks in a genetic sense.

A second way of approaching the problem is to examine the vital parameters of growth, recruitment, and mortality within and between stocks. As noted in Chapter 1, there were three fisheries on the Downs stock in the southern North Sea — the East Anglian fishery in the Southern Bight, the Boulogne fishery on the spawning grounds in the Straits of Dover and in the eastern Channel, and the Belgian fishery on spent herring in the eastern Channel and in the Southern Bight. As indicated from the l_1 studies, there were persistent differences in growth rate between the three spawning groups, but none were found within the stock exploited by the three different fisheries in the southern North Sea. Common mortalities were established by comparing percentage age distributions between fisheries within the same stock for long periods of years, and differences in mortality were shown when such distributions were compared between stocks.

The relationship may be expressed formally:

$$\ln y = a''_1 + (Z_y/Z_x) \ln x, \tag{5}$$

where y is the age-group percentage in the first fishery,

x is the corresponding age-group percentage in the second fishery,

Z_y is the instantaneous total mortality coefficient in the first fishery,

Z_x is the instantaneous total mortality coefficient in the second fishery, and

a''_1 is a constant.

When $Z_y/Z_x = 1$, there is no need to transform the data to logarithms. When $Z_y/Z_x \neq 1$, the logarithmic transformation yields a linear regression. For data series of thirty years or so, age-class percentages in one fishery were plotted on the corresponding percentage in another. The regressions through the origin did not differ from the bisectors, and we conclude that the mortality rates were the same in the three fisheries on the Downs stock. The same technique showed that the mortality rate on the Dogger stock was about half that on the Downs stock. If the mortality rates are the same between percentage age distributions, the percentages of recruits (i.e., three- or four-year-olds) in one fishery, plotted on those in another, for long data series, shows the correlation between year classes (if the mortality rates were higher in one set of samples, the percentages of recruiting year classes would also be higher, even if their true abundances were the same). The year classes found in all three fisheries on the Downs stock were highly correlated, indicating that they shared common year classes. In contrast, the total recruitment to the

Buchan, Dogger, and Downs stocks shows no correlation between any of the three in any combination. Total recruitment is calculated from

$$R_3 = n_3 + [n_4 - n_3 \exp(-Zt)] \exp(Mt)$$
$$+ [n_5 - n_4 \exp(-Zt)] \exp(2Mt), \tag{6}$$

where n_3 is the catch per effort, or stock density, of three-year-old recruits, the first recruiting age group,

n_4 and n_5 are catches per effort, or stock density, of four- and five-year-old fish, respectively,

Z is the total mortality rate,

t is time, one year,

M is the instantaneous coefficient of natural mortality (estimated as 0.2 in Cushing and Bridger, 1966), and

R_3 is the stock density of three-year-old fish that will recruit to the fishery at three, four, and five years of age, i.e., total recruitment.

The lack of correlation in total recruitment between the three spawning groups in the North Sea (Cushing and Bridger, 1966) is in sharp contrast with the strong year-class correlation between the three Downs fisheries in the Southern Bight and eastern Channel.

Within the Southern Bight, there are three fisheries with common year classes, common mortality, and a common growth rate. The three vital parameters in the Southern Bight fisheries exploiting the Downs stock are distinct from those in the fisheries exploiting the Buchan and Dogger stocks. It is possible that there are subpopulations within the Southern Bight group. There are two subsidiary spawning grounds in the Downs group, the first at Sandettié in the Straits of Dover and the second at Ailly in the eastern Channel (Ancellin and Nédelèc, 1959). The vertebral count at Sandettié averages 56.50–56.60 (Ancellin, 1956) whereas that at Ailly averages 56.60–56.70 (Ancellin, 1956), which supports the small quantity of genetic evidence cited above. It is a very small difference compared with that between the American and European eels, but it is consistent and persistent for a long period. Le Gall (1935) made many body measurements on samples of fish from the two spawning grounds, but found no differences except in vertebral count. He also compared the percentages of age-class distribution for a seven-year period between the two grounds and found them to be the same. The vital parameters that are common to the two subsidiary spawning grounds suggest that the single Downs stock has two spawning grounds. The difference in vertebral count between them suggests that they segregate to each ground.

The three groups mix on the feeding grounds in the northern and central North Sea, and appear to segregate to spawning grounds which are very localized. The spawning grounds are very small but are nearly always in the same place. That near the Sandettié Light Vessel is about 2 miles long and 500 yards wide (Bolster and Bridger, 1957), and the size of all others in the North

62   FISHERIES BIOLOGY

Sea appears to be of the same order. There is no evidence that these grounds
have shifted since trawlers discovered them between 1930 and 1950. The fact
that the herring return to them within a week or so of the same dates each year
suggests that they return to the ground of their first spawning. In support of
this conclusion is the further evidence that l_1 distributions remain the same
throughout the life of a year class and that vital parameters are homogeneous.
It is possible that the herring return to their native grounds, but the evidence is
lacking.

The Isolation of Fish Stocks as Shown by Tagging Experiments

Thompson and Herrington (1930) separated the two stocks of Pacific
halibut by means of a particular form of tagging experiment off British Columbia. Figure 22 (Chapter 2) shows the number recovered as a function of
distance from the point of liberation in three successive seasons. Very few fish
were recaptured more than 40 miles away. Numbers decay with distance,
irrespective of direction, as if the fish had diffused away from the point of
liberation. The southern stock off British Columbia spreads up to 40 miles,
and that off Alaska 1000 miles distant, spreads hundreds of miles, yet the
chance of mixture between the two is practically nonexistent. On the basis of
these observations, and a single meristic difference, Thompson and Herrington treated the two stocks separately.

Figure 31 shows the transpacific migration of albacore (Otsu, 1960), which
takes a year or so. This simple observation suggests that the stock of albacore
is contained within the North Pacific anticyclone. This does not mean that the
fish must drift round the circuit in the Kuroshio, Kuroshio extension, California current, and North Equatorial current, but that they may make use of parts
of that system during their life history, as immature albacore are found in the
Kuroshio extension (Suda, 1963). A detailed model of albacore migration of
different ages in the North Pacific ocean was developed by Otsu and Uchida

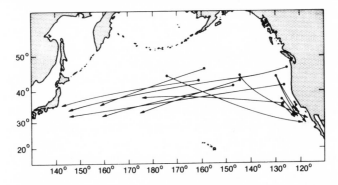

Figure 31. The transpacific migration of the albacore. The point of release is shown by a circle
and the point of recapture by an arrow. Adapted from Otsu, 1960.

(1963). Rothschild and Yong (1970) extended the study with large quantities of material from the Japanese long-line fishery across the North Pacific ocean. The original tagging experiment set the scale on which the subsequent models were developed. Nakamura (1969) summarizes similar information on other species of tuna.

A large tagging experiment on the British Columbian herring (*Clupea harengus pallasi* Valenciennes) was made with internal tags between 1936 and 1956 on spawning or spent fish. The percentages recovered from the ground of tagging on the Upper West and Lower West stocks (west of Vancouver Island) are shown in Table 4. In the Upper West stock the proportion recov-

Table 4. Proportion of tagged British Columbian herring recovered on the Upper West and Lower West areas, and proportion of Lower West fish recovered in the Upper West area (Harden Jones, 1968)

Years after tagging	Recovered on Upper West	Lower West recovered in Upper West	Recovered on Lower West
1	75.7%	13.7%	81.0%
2	82.7	23.9	68.8
3	80.3	24.3	64.5
4	82.3	26.3	61.4
5	87.1	63.1	31.6
6	94.1	83.0	16.7

ered increases with years after tagging, but that in the Lower West stock decreases. The proportion of Lower West fish recovered at Upper West increases sharply with age. Thus, with age, Lower West spawners move to the Upper West spawning ground, which shows that exchange between spawning grounds can take place in herring. A similar analysis suggests that fish from the Lower East and Upper East grounds (east of Vancouver Island) move with age to the Middle East grounds. There appears to be little exchange between the east and west coasts of Vancouver Island, and the results suggest tentatively that there are two stocks, one on each side of the island.

Figure 32 gives the results of a tagging experiment on a much smaller scale (de Veen, 1962). It shows the recaptures of plaice on their spawning grounds where they had been tagged and released a year before. None had strayed from one spawning ground (off Flamborough Head); from the other, 4 out of 56 had moved to one close by (the group south of the Dogger Bank). Fish that mixed on their feeding grounds segregated to their spawning grounds, which might suggest that distinct populations originated on each. However, recent evidence (Purdom, Thompson, and Dando, 1976) shows no genetic differences between the spawning groups in the southern North Sea, from which we may conclude that all three groups should be sampled together in any study

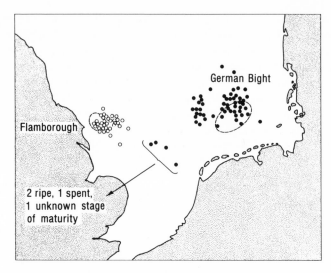

Figure 32. Recaptures of plaice on the spawning grounds where they had been tagged and released a year before. With the exception of four strays from the German Bight, all were recaptured on the grounds where they were released. Adapted from de Veen, 1962.

of recruitment and parent stock. The tagging experiment nevertheless yields an interesting result, that the stray between spawning groups, or stocklets, is low; Li (1955) suggested that most variability would be obtained from a number of spawning groups between which there is little interchange and within which mating is randomly distributed.

Tagging experiments yield results of considerable value in establishing the unity of a stock, but they are only useful if fishing vessels recover the tags across the whole range of migration. In Chapter 2, methods for analyzing the mean direction and distance traveled of tagged fish are given. Those methods may be extended to estimate the "center of gravity" of such distributions and their seasonal changes, which would be of great value in unit stock studies.

The Use of Genetics in Studies of Unit Stocks

Although the study of serological differences in fishes dates back to the early 1950s (e.g., J. E. Cushing and Sprague, 1953), the application of such techniques to the problem of unit stocks is due to Marr (1957). His paper on the subpopulations of the Pacific sardine drew attention to discrete spawning groups off California and raised the question of interchange between them; he suggested that tagging experiments which were very costly should not be used until the biochemical techniques had been tried. Under his influence, Sprague and Vrooman (1962) established the existence of three distinct populations in the California sardine with erythrocyte antigens.

Some differences of biochemical origin are genetically determined. If stocks were distinct, isolated, and coherent from generation to generation, one would expect genetic differences to arise, some of which could be detected by biochemical means. Dannevig (1956) has shown that there are marked differences in muscle amino acids between the Skagerrak cod and the Vest Fjord cod. It is possible, however, that the amino-acid composition of muscles is affected by the nature of the food. Hence, the difference between the Skagerrak and Vest Fjord cod might only be a phenotypic one caused by different diets. It is well known that many antigenic properties are genetically determined in human blood groups. Differences in antigenic properties have been found between stocks of goldfish, *Carassius auratus* Linnaeus (Hildeman, 1956), tuna (J. E. Cushing, 1956), salmon (Ridgway, J. E. Cushing, and Durall, 1958), herring (Sindermann and Mairs, 1959), and of redfish, *Sebastes marinus* Linnaeus, *S. viviparus* Krøyer, and *S. mentella* Travin (Sindermann, 1961). Although antigenic differences have been established between fish populations, the methods are expensive, yielding little information for the labor spent. Sick (1962) introduced zone electrophoresis for the analysis of blood proteins which are genetically determined. The hemoglobins migrate to different degrees in an electric field and so can be separated. The most successful application of the electrophoresis of blood proteins is that on the cod stocks of the North Atlantic by Jamieson and his colleagues. Figure 33 shows the distribution of two alleles, the hemoglobin HbI and the transferrin Tf^c (de Ligny, 1969). With the transferrins, the stocks in the North Sea, Baltic, Faeroe Islands, West Greenland, and Newfoundland are distinct from each other, as tested by simple (2×2) contingency tests. With the hemoglobin allele, the stocks at Iceland, Faeroe Islands, and West Greenland are distinct from that in the North Sea. An interesting point is that the stocks of cod in the North Sea appear to be genetically uniform despite the fact that there are three discrete spawning grounds between which the exchange of tagged fish is relatively low (Bedford, 1966). Graham (1933) wrote that there was one cod stock in the North Sea, primarily with comparison of year-class fluctuations there, with neighboring areas.

The homozygotes *a* and *b* and their heterozygotes *ab* are distributed as ($a^2 + 2ab + b^2$), the Hardy-Weinberg law; if the samples are so distributed, the characters are assumed to be alleles. Jamieson and Turner (1979) established 13 transferrin alleles and 2 hemoglobin alleles in the cod stocks of the North Atlantic. The proportions of alleles can be allocated in ($2 \times n$) tables and chi-square tests used to estimate the genetic distinctness of the stocks. From such estimates, the chance of mixture between the major cod stocks in the North Atlantic is as low as 1 in 10^4. There are, however, some anomalies; for example, the stocks at Iceland that spawn in February and March may include a distinct component that spawns in April, which could comprise the immigrants from West Greenland (Jamieson and Jonsson, 1971). Another interest-

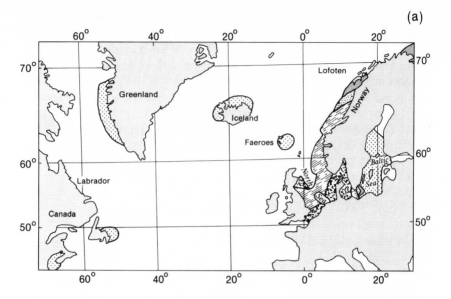

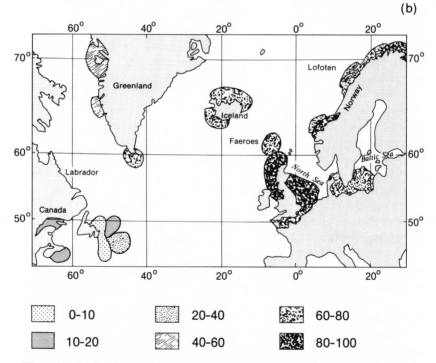

Figure 33. The distribution of two blood alleles in the cod stocks of the North Atlantic: (*a*) distribution of hemoglobin, HbI; (*b*) distribution of transferrin, Tf^c. Adapted from de Ligny, 1969.

ing example was reported by Møller (1968), who showed that genetically distinct groups of cod, coastal cod, and Arcto-Norwegian cod spawn at the same time in the Vest Fjord. Jamieson and Jones (1967) separated the cod stocks on Faeroe Bank and Faeroe Plateau in two successive years on the basis of a single transferrin allele.

Jamieson and Turner (1979) made a more detailed analysis of the cod stocks at eleven positions in the Northwest Atlantic and at five positions off West Greenland. A long series of chi-square tests suggested that there were two distinct groups off West Greenland and four off North America (Flemish Cap, Sable Island, Hamilton Inlet to Halifax, and Georges Bank). Their results may be conveniently expressed in genetic distance, D''. The distance between two populations may be represented as a vector in which gene frequencies, x_i, y_i, specify projections by dimensions, so geometrical principles may be used, Rogers's (1972) estimate is based on Pythagoras and so the maximum distance is $2^{1/2}$; the average is divided by $2^{1/2}$ to yield estimates between zero (where populations are identical) and unity (where they are quite distinct in all gene frequencies):

$$D'' = \left[\Sigma \, (x_i - y_i)^2/2 \right]^{1/2}. \tag{7}$$

Cavalli-Sforza and Edwards (1967) use a more complex index based on many dimensions. The average index is given by

$$D'' = 4\left[1 - \Sigma \, (x_i y_i)^{1/2} \right]/(n - 1). \tag{8}$$

Nei's (1971, 1972) index is based on statistical considerations only:

$$D'' = -\ln \left[(\Sigma x_i y_i)/(\Sigma x_i^2 \Sigma y_i^2)^{1/2} \right]. \tag{9}$$

Figure 34 shows the genetic distances between the six groups already established in the Northwest Atlantic. When $D'' = 0$, we may assume genetic identity. The figure shows the evolutionary history of the cod in this area as six distinct groups. The genetic distances between each pair of groups as shown in the figure are considerable. The important consequence is that any analysis of the dependence of recruitment on parent stock should be based on samples restricted to each of the six groups.

A considerable amount of work has been done on stocks of other species. With esterases and lactic dehydrogenase (LDH), Ridgway, Cushing, and Durall (1971) have established differences between the herring stocks off the eastern seaboard of the United States and the Maritime Provinces of Canada. The stock on Georges Bank is distinct from those in the Gulf of Maine and off southern Nova Scotia, and the immature fish from Passamaquoddy Bay appear to recruit to the adult stocks off southeast Nova Scotia. Fujino (1970) has discovered nine blood group systems in the skipjack tuna (*Euthynnus pelamis* [Linnaeus]), mainly transferrins and serum esterases. He found that the population in the Pacific and the Atlantic were genetically distinct and that that in the western Pacific could be distinguished from those in the central and eastern ocean. The latter conclusion was confirmed with the use of another blood factor. Much work has been carried out on the Atlantic and Pacific salmon stocks. Those spawning in Scandinavian rivers are distinct from those spawning in British rivers, and indeed, from an analysis of gene frequencies, the salmon from those rivers neighboring the Wye in Wales appear to be distinct from the salmon that spawn in Scotland (Payne, Child, and Forrest, 1971). Møller (1970, 1971) has examined samples of Atlantic salmon from Labrador to Maine and separated a number of spawning groups with transferrins. With three transferrin alleles, Utter, Ames, and Hodgins (1970) distinguished coho salmon in Puget Sound from those in the Columbia River. In the crested blenny (*Hypleurochilus geminatus* [Wood]), Johnson (1971) has demonstrated the existence of a cline in frequency of an LDH allele in Puget Sound with latitude, which in view of earlier evidence might imply that there

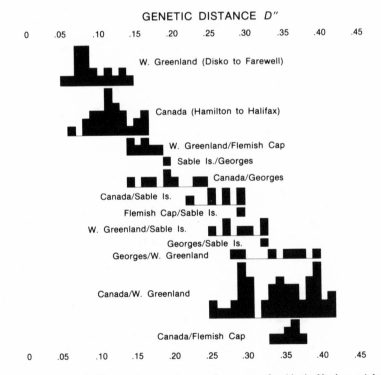

Figure 34. The genetic distances between six spawning groups of cod in the Northwest Atlantic. The genetic distances were estimated with Rogers's (1972) coefficient. The areas indicated are those of the six spawning groups between Georges Bank and West Greenland. Adapted from Jamieson and Turner, 1979.

were a number of isolated stocks distributed from one end of the sound to the other. Iwata (1975), with the use of alleles of tetrazolium oxidase, has separated stocks of walleye pollock (*Theragra chalcogramma* [Pallas]) off Hokkaido, in Japan, and in the Bering Sea.

The work on the genetic isolation of fish stocks is of recent origin, but in the last decade or so has yielded results of considerable interest. Fisheries biologists base their accounts on unit stocks, the vital parameters of which should be homogeneous — particularly recruitment. Various methods were devised to establish the nature of unit stocks, most of which were only partly successful, although it must be admitted that the early biologists, such as W. F. Thompson and Michael Graham, were able to establish them by using common sense, as confirmed by recent genetic studies.

Discussion

The arctic cod is clearly contained within the current systems in which it chooses to live. There is a single but perhaps extensive spawning ground, and there is no problem of mixing with other stocks. The Buchan and Dogger groups of herring have been shown by tagging experiments (Höglund, 1955) to move around the North Sea in the same direction as the main North Sea swirl (Böhnecke, 1922; Tait, 1930, 1937). The three groups of North Sea herring mix on the feeding ground, and probably segregate to their spawning grounds.

There is no positive evidence of coherence between generations in the arctic cod or the North Sea herring. As no other cod stocks live close to that of the arctic cod and as the fish return to their single spawning ground each year, coherence need not be questioned. Yet the Arcto-Norwegian cod stock is genetically distinct from its North Atlantic neighbors. The three groups of North Sea herring mix on the feeding grounds, and so the coherence between generations is justifiably questioned. A limited definition of a unit stock can be used — that year classes recruiting to a given spawning ground appear to return to it until they die. Small groups of herring in the Gulf of Maine are genetically distinct, and one would expect that spawning groups of the same species in the North Sea were also distinct. For Pacific salmon the question has been answered in the demonstration of the parent-stream hypothesis. So far as limited evidence is available, groups of salmon appear to be distinct — i.e., coho salmon from one small group of rivers is distinct from that in another.

Until recently, it was assumed that the North Sea plaice and sole popula-tions were unit stocks, at least in the southern and central North Sea. The Dutch tagging experiments described in Chapter 2 (de Veen, 1961) show that there are three groups of plaice within this area, and that the adults return to the same spawning grounds year after year. In this case, recent evidence has shown that there are no genetic differences between the three spawning groups, which allows us to group them for the purposes of estimating the population parameters. There are, however, genetic differences between plaice in the North Sea and the Irish Sea (Purdom and Wyatt, 1969).

The most important problem to be solved for herring of the North Sea is that of mixing. Rollefsen (1934) has separated the fjord cod in the Vest Fjord from the arctic cod by means of differences in the character of the otoliths, which may well have been confirmed in genetic evidence. The obvious way is to tag fish on the spawning grounds, and the mixing rate is then determined by the proportion of each set of "spawning" tags recovered on the feeding ground. North Sea herring cannot yet be tagged effectively enough to achieve this purpose. If fish were tagged on the spawning grounds of stocks established on a genetic basis, it is possible that the two methods could be combined to analyze the mixing of distinct stocks on a common feeding ground. Then the genetic composition of samples of tagged fish recovered would be of considerable interest.

Despite the promise and power of the genetic methods, meristic characters have on occasion been used to separate unit stocks. When only meristic characters are available, fish of different stocks in a mixed fishery can only be separated by the combination of meristic characters. This is achieved by means of discriminatory functions (Fisher, 1936) or by means of distance functions (Rao, 1952). The techniques are effective so long as the differences in the meristic characters are sufficient. Fukuhara et al. (1962) have classified red salmon of Asian and American origin in the open Pacific. The error of misclassification was about 23 percent, with seven meristic characters combined. Margolis et al. (1966) have used the same method for the sockeye in the open ocean, in combination with tagging observations, scale measurements, and the distribution of parasites. The error of misclassification was lower, and the discriminatory analysis, together with the equally powerful method of scale analysis, provided a clear separation of sockeye stocks in the North Pacific.

The ideal unit stock has a single spawning ground to which the adults return year after year. It is contained within one or more current systems used by the stock to maintain it in the same geographical area. The migration circuit is the means by which this delimitation is achieved and by which the stock maintains its coherence from generation to generation. When mixture takes place, coherence needs to be shown and cannot be assumed.

4
Growth within the Trophic Structure

Four aspects of growth that are apparently quite disconnected from each other are described in this chapter. The first is a study of those simple growth equations needed for the statistical purposes of stock assessment and of more developed growth equations which summarize much carefully gathered information on the physiology of food conversion and on the physics of swimming. Much of the strictly physiological work started with Brown's (1946a, 1946b) experiments on the efficiency of food conversion in brown trout (*Salmo trutta* Linnaeus). Winberg's (1956) estimations of total metabolism from measurements of oxygen consumption by animals of different weights represent another physiological foundation. The second aspect of growth is a comparative study, in groups of fish species, of the constants in the von Bertalanffy (1934) equation. There is a large quantity of data which leads to some broad generalizations: that bigger fish tend to live longer, that short-lived fish grow quickly, and that a fish usually matures at a fixed proportion of its asymptotic weight. The third aspect includes experimental and theoretical studies on feeding behavior. Those studies form the basis of what is known of the predator-prey relationships in fishes. The experimental analysis in terms of behavior is based on the close observation of one fish and the food it takes in rather short time periods. Food intake depends on the quantity of food in the water, its patchiness, and on the sizes of individual organisms. Models of predation are elaborated on the basis of the precise experimental observations of such factors. The fourth and last aspect of growth is discussed as the role of growth in population dynamics, particularly with respect to the density-dependent processes that occur in the sea.

The main problem is to link the growth constants adduced from statistical observations to the physiological material from experimental work and to the behavioral and ecological factors as observed. The synthesis is often made today in the form of brief mathematical models which can be tested either

71

experimentally or by observation. In this way the natural history of a fish population can acquire quantitative expression.

Simple Growth Equations

Simple growth equations are used to graduate weight or length measurements in age. Gray (1926) pointed out that the constants so derived have no exact biological meaning. The processes of growth were described briefly but exactly by Medawar (1945): weight increases with age in a multiplicative way; the rate of growth is constant, but with age the specific growth rate, G', declines,

$$G' = \{[1/(t_1 - t_0)] \ln (W_1/W_0)\}, \tag{10}$$

where W_0 is weight at time t_0, and W_1 is that at time t_1.
Figure 35 shows the growth of plaice in log weight on age from larval life to old age, and so it is a curve of specific growth. The slope represents the specific growth rate, and it clearly declines with age. From the age of

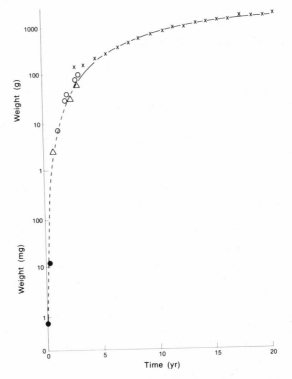

Figure 35. The growth in weight of plaice expressed in logarithms during twenty years of life. Weights of yolk-sac and later larvae are shown by solid circles, weights of 0-group fish by triangles, weights of juveniles by open circles, and weights of adults by X's. Adapted from Cushing, 1975b.

metamorphosis onward, the observations are fitted by the von Bertalanffy (1934) equation, but those of larval weight are not fitted by the curve. The figure also shows that from the age of four, the age at maturity, the plaice grows by more than an order of magnitude before old age. The growth of fishes differs from that of terrestrial animals in that the specific growth rate does not decrease so sharply at the age of maturity. It was for this reason that Bidder (1925) suggested that fish were potentially immortal. The age of maturity can sometimes be detected by a spawning ring on scale or otolith, but there is no evidence of change at that age in the regular decline of the specific growth rate.

The von Bertalanffy equation fits many observations on the growth of fishes, crustacea, or molluscs, and the constants can be incorporated readily into stock assessment models. The simplest derivation is that of Gulland (1969):

$$dl/dt = K(L_\infty - l),\tag{11}$$

where l is the length of fish in cm,
   L_∞ is the asymptotic length of fish in cm,
   K is the rate at which length tends toward the asymptote, and
   t is time.
Integrating,

$$L_\infty - l = c \exp(-Kt), \text{ or } l = L_\infty - c \exp(-Kt),\tag{12}$$

where c is the constant of integration.
Let $l = 0$ at t_0, then $L_\infty - c \exp(-Kt_0) = 0$,

$$\therefore c = L_\infty \exp(Kt_0);\tag{13}$$

$$\therefore l_t = L_\infty \{1 - \exp[-K(t_1 - t_0)]\}\tag{14}$$

From estimates of length at age, the annual increment of length $(l_{t+1} - l_t)$ is plotted on length l_t, where l_{t+1} is length at age $t+1$ and l_t that at age t,

$$l_{t+1} - l_t = (L_\infty - l_t)[1 - \exp(-K)];\tag{15}$$

the slope of the line is $[1 - \exp(-K)]$ and the intercept on the abscissa is L_∞. This expression establishes the constants, but it also expresses the decline of the specific growth rate with age. In the Ford-Walford plot (Walford, 1946), l_{t+1} is plotted on l_t, the slope of which is $(\exp - K)$, and L_∞ is given by the intercept on the bisector. Gulland (1969) gives methods for use when the intervals of time are not equal, as for example in the observations on the lengths of tagged fish. Ricker (1975) discusses certain biases in estimation.

Because weight varies as l^3, Equation (14) can also be expressed in weight:

$$W_t = W_\infty \{1 - \exp[-K(t_1 - t_0)]\}^3.\tag{16}$$

An important point is that the constant K is the same in both equations.

There are many growth equations, most of which may be expressed in a general form (Richards, 1959; Gulland, 1969):

$$W_t^{(1-m')} = W_\infty^{(1-m')}[1 - d' \exp(-Kt)], \qquad (17)$$

where d' is a constant.
The von Bertalanffy equation is given by $m' = 2/3$, the autocatalytic by $m' = 2$, the monomolecular by $m' = 0$, and the Gompertz by $m' \to 1$. Allen (1969) used an analogous generalization from Taylor (1962). Of the equations, the von Bertalanffy equation is the most convenient, but not all observations on the growth of fishes are fitted by it, and the others are available. But Parker and Larkin (1959) formulated an equation which expresses different "stanzas" in different environments:

$$dW/dt = k'W; \; W^{(1-x)} = (1-x)k't + W_0^{(1-x)}, \qquad (18)$$

where k' is a scaling factor and x a power.
Sockeye salmon and steelhead trout migrate from fresh water to the sea where they grow more quickly, which is expressed as a significant increase in the power x. The von Bertalanffy equation demands that growth is isometric with age, and the Parker-Larkin one would express the change in shape at metamorphosis as well as other changes as fishes move from one environment to another.

Fisheries biologists use the simple growth equations to summarize much information on length or weight at age, to compare the growth of different species, or of different stocks, in various environments, and to express the changes of weight and of numbers with age in the form of yield curves as function of fishing mortality or of stock.

Development of More Complex Growth Equations

From earlier work by Ivlev (1961) and Winberg (1956), Warren and Davis (1967) specified the components of growth in fishes:

$$Q'_c - Q'_w = Q'_g + Q'_s + Q'_d + Q'_a, \qquad (19)$$

where Q'_c is the energy in the daily ration r, Q'_w that in excretion, Q'_g that in the weight increment, Q'_s that of standard metabolism, Q'_d that of specific dynamic activity (the energy of transfer from its sources), and Q'_a is that of swimming activity. Such quantities expressed in calories can be put as rates, e.g., $Q'_c/(W^{0.8}t)$, which vary considerably with size, as discussed above in relation to the decline of specific growth rate with age. Winberg (1956) estimated the daily ration and the daily weight gain as proportions of weight, using measurements of oxygen consumption by animals of different weight (Table 5).
Ivlev had expressed the efficiency of growth in two coefficients, that of feeding, $K'_1 = \Delta W/(r\Delta t)$, and that of absorption, $K'_2 = \Delta W/(a^*r\Delta t)$, where

Table 5. Calculated daily ration and weight increment as percentage
of weight of fish (after Winberg, 1956)

Wt of fish in g wet wt	Daily ration as percentage of wt	Daily weight gain as percentage of wt
0.001	37.6–89.5	1.50–43.0
0.1	14.9–35.4	0.59–17.0
10.0	6.0–14.4	0.24–6.9
1000.0	2.4–5.6	0.09–2.7

$a*$ is the proportion of food absorbed and where r is daily ration in weight. Paloheimo and Dickie (1965, 1966a, 1966b) proceeded from Ivlev's first coefficient as follows:

$$K'_1 = \exp(-a'' - b'r) \text{ (or ``K-line'')}, \tag{20}$$

where a'' and b' are constants.

$$T' = \alpha* W^\delta \tag{21}$$

where T', the metabolic rate, is ml O_2/g body weight, and $\delta = 0.8$; Winberg (1956) had established that $T' = 0.3W^{0.8}$ ml/h at 20° C.

$$\Delta W/\Delta t = r - T', \tag{22}$$

$$\therefore T' = r[1 - \exp(-a'' - b'r)]. \tag{23}$$

The "K-line" subsumed much published information in terms of growth efficiency, but it takes no account of losses in weight or of feeding at a maintenance level.

As will be shown below, fish do not maximize their growth efficiency (in contrast to Paloheimo and Dickie's approach, which suggests that they do), but their growth rates, and so they will get the ration they need. However, growth efficiency is of considerable importance in exploitation in that fish should be caught when their growth efficiency has become reduced, i.e., when they are big and have converted food into fish flesh.

The second coefficient can be estimated experimentally, for example by Birkett (1969):

$$\Delta W = \xi'(r - A_1), \tag{24}$$

where ξ' is the efficiency of food conversion in units of nitrogen,

A_1 is the maintenance threshold in units of nitrogen, energy, etc.

Edwards, Finlayson, and Steele (1969) normalized such quantities to units of $W^{0.7-0.8}$. Birkett derived a useful ratio, $(1 - \xi')/\xi'$, the metabolic cost of growth. Such expressions can be estimated under different environmental conditions and experimental regimes to analyze the physiology of growth.

However, to understand the observed growth of fishes we need to express the energy spent in swimming. D'Arcy Thompson (1917) had noted that the size of terrestrial animals was limited by the ratio of potential work (αl^4;

weight × distance) to actual work done ($\alpha\ l^3$, as function of the mass of muscles). But by evading gravity, aquatic animals — such as fishes — are both larger and smaller than terrestrial ones. When they swim, they have to overcome drag: Froude's law states that work done varies as the product of drag ($\alpha\ l^2$) and the square of the velocity, so

$$V^2\ \alpha(l^3/l^2)\ \alpha\ l,\ \therefore\ V\ \alpha\ l^{1/2}.$$

Gray (1968), with Bainbridge's (1960) results and a more rigorous derivation, concluded that $V\ \alpha\ l^{0.6}$.

Greer Walker (1970) showed that drag increases with speed much more in big fish than in little ones, and so the relative cruising speed (lengths/sec) decreases with length. Much experimental material has been published on the swimming speeds of fishes (Webb, 1975), particularly by Brett (1964, 1965) and his school on the oxygen consumption of sockeye salmon as they swam for long periods.

Many physicists and biologists have modeled the transfer of energy from fish to the water. Lighthill (1971) stated that as soon as any segment of the body moves, a reaction force is generated proportional to the velocity of that segment and the virtual mass of the reacting water. Then thrust is proportional to the rate at which momentum is shed into the wake. In general, thrust power is proportional to weight, Reynolds number (inertial force/viscous force) increases with length, and the drag coefficient decreases with length. Thus the size of fishes is limited in the end by Reynolds number, and bigger fish tend to swim more slowly than they could. An interesting point is that the energy cost/km is least at low swimming speeds at low temperatures, so fish should migrate in deep water or at cool seasons. This brief summary is taken from Webb (1975).

Kerr (1971a, 1971b) examined growth efficiency as a function of swimming and searching and established a set of "K-lines" for different weights and densities. With the same system and a series of food densities in the life history, Kerr (1971c) was able to fit a growth curve to the observed weights of Lake Opeongo trout; the metabolic rate that maximizes growth efficiency/ration was estimated in the model, from which the daily ration and then growth were determined. However, Ware (1975) adopted a different approach, using Holling's formulation of food intake with search behavior and an analogous cost of swimming. With Ivlev's original data on experiments on bleak, he showed that the fish maximized growth rate and not growth efficiency. In other words, the demand for food must be satisfied irrespective of the immediate efficiency of obtaining it.

R. Jones (1976a) reduced many concepts to simple empirical equations:

maintenance energy = $0.03W^{0.8}$ exp $0.81T$ k J/day at T°C,   (25)

swimming energy = $0.008W^{0.8}$ exp $(0.081T + 0.76V)$ k J/day at T°C, (26)

where V is swimming speed in lengths/sec.

Combining the two equations, he derived an empirical one for gadoids:

$$\Delta W = 0.79 \, [H'W^{-0.15} - 0.008W^{0.65} \exp \, (0.081T + 0.76V)] \quad (27)$$

g/day at T°C, where H' is the increment of food.

With the type of experiment done by Brett and Birkett and the fundamental Winberg power law on oxygen consumption per unit weight, Jones has elaborated an expression for growth in weight. In present population studies an expression in weight is probably more useful than one in energy.

Comparative Study of Growth

A lake sturgeon (*Acipenser fulvescens* Rafinesque) has been found with 151 rings on its scales, but such fish from Lake Winnebago normally live to about 45 (Probst and Cooper, 1954). The greatest age of *Acipenser stellatus* Pall. from the Caspian Sea ranges from 16 to 31 years (Nikolskii, 1969). A freshwater atherinid, *Labidesthes sicculus* (Cope), spawns at the age of one year and dies two or three months later (Hubbs, 1921). A North Sea herring lives about twelve years (Hodgson, 1925), but the larger Norwegian herring lives to the age of 23 (Lea, 1930). Cod in the Barents Sea reach the age of 25 (Rollefson, 1954), but in the Sea of Japan they are old at the age of 12 (Nikolskii, 1969). Plaice have been found with 37 rings on their otoliths (Cushing, 1975a), whereas sprats live for five years and anchovies for three (Nikolskii, 1969). Nikolskii (1969) has constructed a distribution of maximum age for 104 species and one of greatest length for 177 species; the mean maximum age was 16.5 ± 13.5. The origin of differences in longevity is evolutionary and inaccessible, but some are illuminated by a comparison of growth constants.

Figure 36 shows the relationship between the oldest recorded age and the asymptotic length, L_∞, for the Clupeoidei and the Pleuronectoidei (Beverton

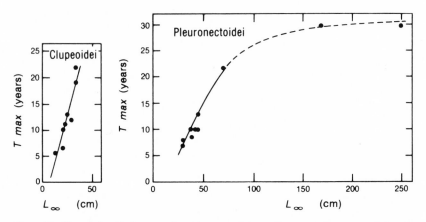

Figure 36. The relationship between the oldest recorded age, T_{max}, and the asymptotic length, L_∞, for the Clupeoidei and the Pleuronectoidei. Adapted from Beverton and Holt, 1959.

and Holt, 1959). Bigger fish live longer, and it can be shown that cod-like fishes are bigger for their age than herring-like fishes. In Figure 37 a similar dependence is given (for the Clupeoidei and the Gadiformes) of natural mortality, M, on the rate at which growth to the asymptotic length decreases, K; fish that grow quickly die young. If the two figures are compared, K is inversely related to the maximum age, and therefore the correlation between M and K should not be unexpected. It really is stated in the von Bertalanffy equation.

Beverton (1963) established a relationship between L_m (the length at first reaching maturity) and L_∞ for a number of clupeid families. The bigger the

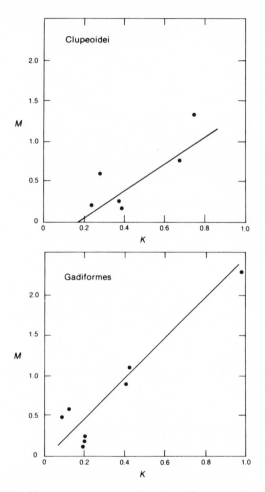

Figure 37. The relationship between natural mortality, M, and the rate at which fish grew to their asymptotic length, K, for the Clupeoidei and the Gadiformes. Adapted from Beverton and Holt, 1959.

fish is, the bigger it is on first reaching maturity. In other words, L_m/L_∞ is constant within a family. In many clupeid families, the two ratios M/K and L_m/L_∞ tend to be inversely related, perhaps because K and L_∞ are inversely related. Hence it follows that when the reproductive load is high (i.e., high L_m/L_∞), the ratio M/K is low, as in the Clupeoidei (1.4), but in the more reproductively self-sufficient Gadiformes it is higher, (2.2). The quantitative relationships are less valuable than the principles that emerge. Final size represents the cumulative difference between growth and death throughout the life history. A large fish lives for a long time and its natural mortality (in adult life) is low, which means that predation has been avoided, perhaps merely because the animal is large; all animals in the sea climb through the Eltonian pyramid, and the larger survivors are not very numerous. A small fish does not live long; it suffers a high natural mortality and a high reproductive load. Fish produce eggs in proportion to their weight, and so the numerical losses in juvenile life also depend on weight. Hence the difference between growth and death in larval and juvenile life must be greater in big fish than in small fish.

Predator and Prey

Ivlev (1961) kept freshwater fish in small tanks for 1½–2 hrs and estimated the quantity of food eaten (1) from the quantity left over from the quantity fed to the fish, (2) from the quantity seen by an observer to be eaten, and (3) from gut contents, the fish having been starved for 18–20 hrs before each experiment. Figure 38 shows the effect of food concentration, p', upon the daily ration, r; the observations are fitted by the equation:

$$r = r_{max} [1 - \exp(-k''p')], \tag{28}$$

where r_{max} is the maximum ration and k'' the feeding coefficient.

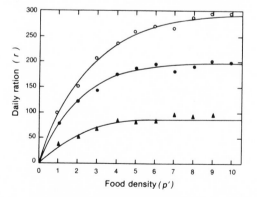

Figure 38. The effect of concentration of food, p', in mg/cm² on the daily ration, r, in mg for carp (*Cyprinus carpio* Linnaeus) feeding on bream (*Abramis brama* Linnaeus) roe (open circles), roach (*Leuciscus rutilus* Fleming) feeding on chironomid larvae (closed circles), and bleak (*Alburnus alburnus* Linnaeus) feeding on *Daphnia* (triangles). Adapted from Ivlev, 1961.

80   FISHERIES BIOLOGY

Savage (1937) examined the gut contents from 100 North Sea herring in each fortnight from May through September for five years. With information on digestion rate (Battle, 1935), the numbers of *Calanus* found in the gut can be converted into quantities eaten per day. The number of individuals in the gut each day represents the number of successful attacks made, and the daily weight in the gut is a measure of the effectiveness of feeding. Figure 39 shows the relationship between the number of attacks made daily (averaged for fortnightly periods) and the weight of food for the year 1931; such observations and those from the other four years could well be fitted by an Ivlev curve.

More recently, Holling (1965) has analyzed predation by splitting the processes into components of time: the time to capture and eat a prey, t'_c, is divided in four parts: (1) the digestive pause, t_α, during which no attack occurs; (2) the time spent searching, t_s; (3) the time spent in pursuit, t_p; and (4) the time spent in handling and eating. Hunger, H', was expressed as

$$H' = H'_k + (H'_0 - H'_k) [\exp - f't)], \tag{29}$$

where H' is g food needed to satiate;

H'_k is the maximum capacity of gut in g,

H'_0 is the fullness of gut needed to satiate in g,

f' is the specific rate at which the gut is emptied.

The time to search is given by:

$$t_s = \frac{1}{2V_r\beta'' (H'_0 - H'_{t_0}) N_0 S_s} - \frac{\pi\beta'' (H'_0 - H'_{t_0})}{2V_r}, \tag{30}$$

where V_r is the difference between the speed of predator and prey in cm/sec,

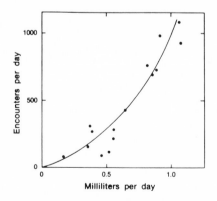

Figure 39. Number of attacks or encounters made daily (averaged for fortnightly periods) by herring, plotted on the weight of food in the gut, in the year 1931. Adapted from Cushing, 1964, after data in Savage, 1937.

β'' is a constant determined by the shape of the search field and the rate at which it expands with hunger,

H'_{t_0} is the quantity of food needed to satiate when search starts, in g,

N_0 is the prey density,

S_s is the strike success.

The time spent in pursuit is

$$t_p = (\beta''/V''_p)\,(H'_l - H'_{t_l}), \tag{31}$$

where V''_p is the predator's attack speed in cm/sec,

H'_l is the quantity of food needed to satiate when the prey is captured in g,

H'_{t_l} is the attack threshold when the prey is captured (as quantity of food in the gut).

Holling's method reduces many different physiological and ecological processes to units of time, which simplifies biological complexity to mathematical terms. It represents a considerable advance on the somewhat oversimplified Ivlev equation, but requires a higher quality of observation.

The Ivlev curve can, however, be used to estimate the effect of patchiness by means of an index of aggregation, $\zeta = [\Sigma\,(v^2/\bar{p})]^{1/2}$, where v is the deviation in density from the mean of a series in space and $\bar{p}$ the mean food density. An adaptation of the Ivlev equation may be written $r = r_{max}\,[1 - \exp(-\chi\bar{p})]$, where χ is the concentration coefficient. Figure 40 shows the daily ration as a percentage of the maximum ration plotted on ζ, the index of aggregation. In other words, as food at a constant mean density becomes more patchy, so the daily ration increases. The relation is expressed by

$$r = \gamma + (r_{max} - \bar{p})\,[1 - \exp(-\chi\zeta)], \tag{32}$$

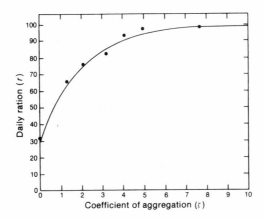

Figure 40. The daily ration, r, as a percentage of a maximum ration, γ, plotted on the index of aggregation, ζ. In other words, the daily ration increases with patchiness. Adapted from Ivlev, 1961.

where γ is the daily ration at zero aggregation, as a percentage of the maximum ration. This curve is drawn to the data in Figure 40. Combining the two approaches, Ivlev showed that when carp (*Cyprinus carpio* Linnaeus) were feeding on benthos in the Volga, the patchiness of the food was as important as its density in governing the magnitude of the daily ration (as percentage of the maximum).

Patchiness is the rule in the sea from the small scales in chlorophyll and temperature dominated by turbulence (Platt, 1975) to the large patches perhaps 50 or 100 km across, sustained against diffusion by variations in reproduction and predation in space and time. On a small scale, patchiness promotes efficient feeding, since the distance between patches may be considerably less than the summed distance between a number of evenly distributed individuals. On the large scale, pelagic fish aggregate onto patches of their food because they slow down to eat and hence gather where food is abundant. Such processes were shown and analyzed by Cushing (1952, 1955). R. Jones and Hall (1973) have shown that larval haddock need more than an average density in order to survive. Even larval haddock might aggregate transiently on small and transitory patches of their food, but such processes might be hard to establish, as they lie outside the range of visual observation by a diver and within the range of most of our sampling instruments. The first feeding larvae of anchovy gather on thin layers of the dinoflagellate *Gymnodinium*, close to the shore off California (Lasker, 1975). Sperm whales (*Physeter macrocephalus* Linnaeus) gather in the equatorial upwelling to feed on patches of food that might be dynamically sustained for thousands of miles (Townsend, 1935). The stock of Atlanto-Scandian herring lay in autumn southeast of Iceland in large shoals which were discovered and reduced by purse seiners, so extinguishing the fishery. The principle might be expressed in an opposed way, that the animals must and do exploit the patches where their food is more abundant.

There is an analogy in patchiness with shoaling fish, developed by Brock and Riffenburgh (1960). They showed that the numbers of prey eaten are high when there are many of them or when the sighting range of the predator is greater than the distance between shoals, which explains why, in general, fish do not shoal at night or in turbid water. Hence there is an advantage in shoaling when predators are not too abundant, which must be the usual case. Fish shoal to minimize predation, but predators find the same advantage in shoaling prey that any patchy food provides. The theory of search in underwater warfare was applied by Olson (1964) and by Saila and Flowers (1969). They showed that more prey came into the predator's range at higher prey speed, so predators are active at dawn and dusk, when prey are active. Conversely, prey should move slowly if they are to survive; Cushing (1976b) has suggested that fish in bigger shoals cruise more slowly than those in small ones. If the predator only looked ahead, it need only swim twice as fast as the prey. Then prey and predator both swim slowly, which agrees with Weihs's (1973) observation on hydrodynamic grounds that fish should cruise at about only one body length/sec.

Cushing (1976b) distinguished predators from filterers. A filterer — such as a herring — cruises all day on its red muscle (Barets, 1961) and feeds all day on more than a thousand *Calanus*, each of which weighs five orders of magnitude less than a herring. A large cod uses its white muscle to attack whiting a few times a day and weighs about a hundred times as much as its prey. The herring can take its very small prey at cruising speed, but the cod must attack at high speed. They can exchange their roles, as for example when herring attack sandeels or when cod filter euphausids. The herring-like fishes shoal readily, but the gadoids do not. We may make the generalization that filterers are shoaling fish that feed at cruising speeds or less.

Ivlev (1961) devised an index of food selection,

$$\eta = (U_i - U_j/(U_i + U_j), \tag{33}$$

where U_i is the percentage of ingredient i, and
  U_j is the percentage of ingredient j.

When η is positive, the predator selects its prey; if negative, it selects against it. Figure 41 shows the index of selection as function of prey size for four predator/prey experiments; it also gives the percentage distribution in weight or length of the prey eaten. The simple conclusion is that there is an optimum

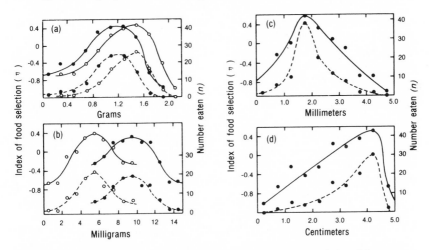

Figure 41. Relationship between the food selection index (solid line) and size of the prey as a percentage distribution in weight or length of the number eaten (dashed line). There is an optimum size of prey for each predator. The four parts of the figure represent: (*a*) pike fed on roach (open circles), perch fed on roach (closed circles); (*b*) carp fed on chironomid larvae (open circles), bream fed on amphipods (closed circles); (*c*) bleak fed on water fleas; and (*d*) larvae of a water beetle (*Macrodytes circumflexus*) fed on young roach. Adapted from Ivlev, 1961.

size of prey for each predator, but the skewed curve in Figure 41 suggests that the largest prey may have had some capacity to escape their predators. Cushing (1964) showed that the greatest quantity of *Calanus* per gut was found at relatively low feeding efficiencies expressed as g/encounter. In other words *Calanus*, the preferred food of the herring, is probably of the optimum size. A similar analysis was made by Slobodkin (1966) of *Daphnia* feeding on algal cells. By dividing the optimum weight of prey by the weight of the predator, Ivlev established rapacity indices for his experimental animals: bleak (*Alburnus alburnus* Linnaeus) 0.00038; bream (*Abramis brama* Linnaeus) 0.0019; carp, 0.13; perch (*Perca fluviatilis* Linnaeus), 0.18; pike (*Esox lucius* Linnaeus), 0.32; and a water beetle of the genus *Macrodytes*, 0.65. Presumably a very active and rapacious fish can afford to be idle for considerable periods, whereas a less active fish has to maintain its feeding more continuously.

A fish's egg weighs a milligram, and a fish grows by six to seven orders of magnitude during its life. Figure 41 suggests that the spread of prey size ranges from half to one and a half times the mean. Ursin (1973), with an extensive analysis of the sizes of food organisms, showed that on average the predator was larger than the prey by two orders of magnitude in weight or by about 4.6 in length. When a prey grows by two orders of magnitude in weight it would pass through the attack fields of six or seven predators, each larger than its predecessor by a factor of two. From such a principle it follows that any death rate due to predation must decline with size of prey because ration/ body weight of the predator decreases with body weight less quickly than does numbers/unit volume. Just as specific growth rate declines with age, so may the specific rate of natural mortality.

The Role of Growth in Population Dynamics

During their lives, fish grow from the egg by four to seven orders of magnitude and lay eggs to the same degree, because fecundity is proportional to weight. Hence if growth were inversely density dependent during adult life, the number of eggs produced would be reduced at high stock. The link between growth and fecundity would provide a powerful mechanism of population control.

Growth has been compared at different densities in lakes and culture ponds (Swingle and Smith, 1943; Johnson, 1965). Backiel and Le Cren (1967) established a clear inverse relationship between specific growth rate and numbers for carp and trout; the differences were considerable. Alm (1959) suggested that carp grew more quickly at low stock with a lower age of maturation than at high stock, which is an additional mechanism of potential population control.

For a single population, the evidence for such changes is scanty. However, Le Cren (1958) has studied the growth rate of perch in Windermere, in northwest England, during a period when its population was reduced by a factor of sixteen. No differences were found in the growth rate of immature animals during their first and second years, but that of older and mature perch increased sharply after the first year of population decline and subsequently. Hence there was enough food for the immature fish that feed on plankton at high stock, but the growth and fecundity of adults must have been limited by food lack. The perch population in Windermere could well have been controlled by density-dependent fecundity, which increased by a factor of 3 for a sixteen-fold reduction in stock (Le Cren, 1965).

Le Cren (1965) carried out a field experiment on the growth and mortality of trout fry in sections in a small stream. Each section was stocked with different quantities of fry, and the survivors were recovered by electric fishing and from fry-proof cages. Figure 42 shows that the growth rate was reduced

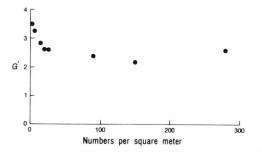

Figure 42. The reduction in specific growth rate (*G'*) with increased density in Le Cren's experiment on the growth and mortality of trout fry. Data from Le Cren, 1965.

with numbers to a minimum. The observations can be fitted by the following equation:

$$\ln [1 - (G'_{max}/G')] = k^*N_0, \qquad (34)$$

where k^* is the rate at which the minimal growth rate is reached in stocking density. Most of the fry that died weighed less than the first feeders, so they had starved. The reduction of growth rate to a minimum suggests a least growth rate for the survivors. The specific growth may depend on numbers in a logarithmic way; Le Cren showed that mortality was also density dependent. If both growth and mortality are density dependent in a logarithmic way, $(G - M) \propto \ln N$, as shown in Figure 43. Growth and mortality appear to be linked in such a way that numbers are reduced to match the density of available food.

The Pacific salmon has been extensively studied, and its growth rate is inversely density dependent in the immature stages of its life history. Foerster (1968) reported an inverse relationship between weight and numbers of one-

year-old sockeye salmon in Cultus Lake, British Columbia, but none in the
Karluk River (southwest Alaska). Mathisen (1969) showed that the growth
rates of both one- and two-year-old smolts and of both second- and third-year
ocean females from Kvichak River are all density dependent; for year classes
that differed by a factor of 60, weights were reduced by 18 to 33 percent.
Similar results were reported by Burgner et al. (1969). The remarkable point
is that the degree of density dependence is less than might be expected from
observations on purely freshwater stocks. It seems unlikely that the stock-
dependent differences in recruitment observed in the Pacific salmon (see
chapter 7) can be effected by such small changes in fecundity; and because
nearly all the fish spawn at the age of four, the differences cannot be effected
by changes in the age of first maturity.

   Because fisheries biologists collected material on adult fishes rather than on
juveniles, density-dependent growth in marine species has not been readily
detected. However, because the growth of juvenile clupeids can be back-
calculated from the scales of adult fish, growth has been shown to be density
dependent in the first year of life but not in later years. The phenomenon was
first demonstrated by Marr (1960) in the California sardine. Iles (1967, 1968)
examined the growth of the East Anglian herring in the year classes 1939–63
and found that growth in the first summer is inversely proportional to the
abundance of larvae hatched in the winter before. No such relationship exists

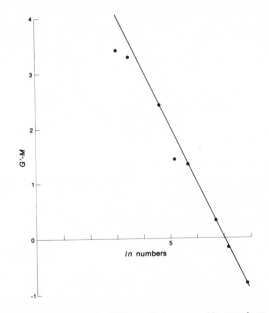

Figure 43. The relationship between the difference between specific growth rate (G') and specific
death rate (M) and the logarithm of density in numbers. Data from Le Cren, 1965.

between the growth increments and stocks by year classes of older fish. Iles correlated the length of succeeding age groups; the coefficients of determination were as follows:

	0/I	I/II	II/III	III/IV	IV/V
r^2	0.144	0.570	0.830	0.904	0.945

Such figures may well reflect the decay of density dependence in growth with age. Lett and Kohler (1976) have shown that the first year's growth of the St. Lawrence herring was inversely related to year class strength, an increment of 1.6 in weight for a decrease in stock by an order of magnitude.

Extensive studies of flatfish have revealed no marked density dependence of growth. Hempel (1955) compared pre- and post-Second World War year classes of plaice at four years of age from eastern North Sea samples and found no relationship between growth increments and density. But Hempel's summary of the growth of all age groups in 1946 showed a marked decrease as compared with pre-war samples, and Beverton and Holt (1957) pointed out that W_∞ was reduced. This result is shown in Figure 44. The transwartime increase in the plaice stock was greater than the variations examined by Hempel. However, my colleague R. C. A. Bannister has shown that W_∞ has declined by year class since 1946 and has not returned to the higher pre-war value as the stock was reduced. In recent years, the sole stock in the North Sea has been reduced considerably, and the weight of the 0-group fish in their first summer doubled during the period (de Veen, 1976). The most remarkable point, however, is that the growth increments of older fish remained constant, but that between three- and four-year-olds increased, so that W_∞ increased. The 50 percent length at maturity increased, but the age of maturation re-

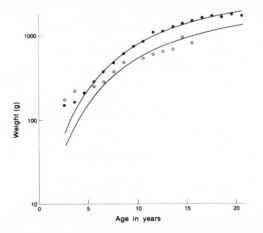

Figure 44. The logarithmic growth curves of plaice before the Second World War (closed circles) and after it, in 1946–70 (open circles). Adapted from Beverton and Holt, 1957, and personal communication from R. C. A. Bannister.

mained constant. Southward (1967) studied the growth of the Pacific halibut from the 1907 year class to that of 1963 from back-calculated measurements from the otoliths of older fish. The growth in the first year is constant, perhaps because the otolith is formed part-way through larval life. Of the increments in length, all are constant with time, except (l_3/l_2) and (l_4/l_3) which increased slowly to the early 1930s, after which it declined, i.e., during the period of stock decline and recovery (see Fig. 45). From tenuous evidence derived from a very large quantity of material, density-dependent growth in flatfish is restricted to young animals, perhaps as old as two or three.

For gadoids, the evidence is scattered. D. S. Raitt (1939) showed that of twelve year classes of North Sea haddock, nine were of about the same abundance, and the fish were all about the same weight. Two year classes were twice as abundant, and the weights of immature fish were reduced by about one-fifth. One year class was five times as strong as the nine, and the immature fish were two-fifths lighter. D. F. S. Raitt (1968) wrote that as the Norway pout *(Trisopterus esmarkii* [Nilsson]) stock varied in the North Sea

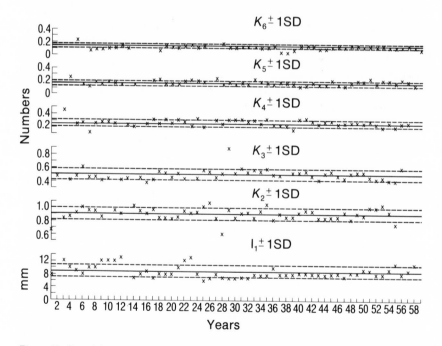

Figure 45. Growth increments (l_1 in the first year, K_2–K_6 in the years 2–6) in the stock of Pacific halibut. There is a possible increase in K_3 until 1931, after which it may have declined; the stock was lowest in 1931. The dotted lines indicate one standard error about the mean. Adapted from Southward, 1967.

by about an order of magnitude, so the predominant I-group fish varied by a factor of 1.67 in weight. Length at maturation remained the same, but the age of first maturity decreased with increased abundance. It is possible, but not fully shown because the data set was small, that the Norway pout responded to population change with density-dependent fecundity. Cushing and Horwood (1977) showed that the growth of four-year-old immature cod of the Arcto-Norwegian stock was inversely density dependent, but not markedly so.

The conclusion from this somewhat discursive account is that the clear pattern of density-dependent fecundity established in lakes or ponds is not shown in the stocks of Pacific salmon and is somewhat difficult to find in marine fish. Density-dependent growth occurs in juvenile fish, in the first summer for clupeids, in the first two or three years for flatfish, and perhaps in the first three or four years in gadoids.*

One of the important conclusions is that differences in growth rate in the sea are related to available food independently of density. In Figure 46, the mean length of each age group (3–9) of herring in the East Anglian fishery are shown for the periods 1932–39 and 1946–59 (Cushing, 1960). The mean

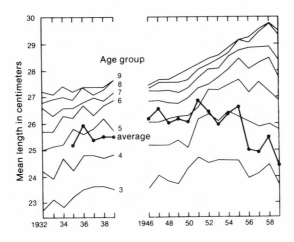

Figure 46. Mean lengths, in cm, of herring of age groups 3–9 in the East Anglian fishery in 1932–39 and 1946–59. The mean length is shown as a bold line; it decreased sharply after 1954, as the older age groups vanished from the fishery. There were increments in length of three-year-old fish between 1949 and 1950, and of four-year-old fish between 1950 and 1951. Adapted from Cushing, 1960.

---

*Recent work by my colleagues R. G. Houghton and S. Flatman has suggested that the growth of adult cod off the northeast coast of England is density dependent.

length of the stock, drawn as a bolder line, declines sharply in later years as older fish disappear from the fishery. Burd and Cushing (1962) related the growth of herring to the number of *Calanus* estimated to be in the area between the English coast and the Dogger Bank from April to September in the periods 1933–38 and 1946–58. The samples were taken near the area in which the immature herring live. Figure 47 shows the relationship between the mean length of the three-year-old herring off East Anglia and the density

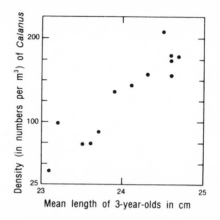

Figure 47. Relationship between the mean length of three-year-old herring at East Anglia and the density of *Calanus*, averaged for the three years of growth. Adapted from Burd and Cushing, 1962.

of *Calanus* in the sea averaged for the three years of the fish's growth. The age of maturation shifted in 1925–28 from just over 5 to about 3½ and in 1950–52 from 3½ to 3. During the period 1952–58, the length at which virgin fish start to mature remained at 21 or 22 cm, whether they were two or three years of age. The changes in length-at-age and in age of maturation were linked to changes in the quantity of *Calanus* available and not to changes in stock density; the latter were considerable during the period of the Second World War and subsequently, and their effect on growth was limited in age to the first summer.

We conclude that in the sea, density-dependent growth can be detected in 0 groups among clupeids and in somewhat older juveniles in flatfish and gadoids. In many of the marine fish examined — herring, Pacific halibut, and North Sea sole — variation in growth independently of stock density was the most pronounced feature observed. Figure 48 expresses the forms of density-dependent growth seen in freshwater fish and in marine fish. In fresh water, the specific growth rate at all ages declines more sharply at high stock than at low stock, and the age of maturation may increase; consequently density-

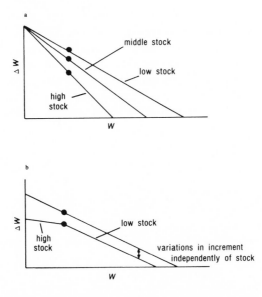

Figure 48. Two forms of density-dependent growth: (*a*) specific growth rate varies with density in adult life, which leads to density-dependent fecundity; (*b*) specific growth rate varies with density only in juvenile life.

dependent fecundity is a possible control mechanism. In the Pacific salmon, the specific growth rate may be reduced at high stock in all juvenile stages, but the age of maturation is not affected. Density-dependent fecundity may occur at high stock because egg production is a function of weight, but this probably cannot account for all the differences in stock-dependent recruitment. In marine fish, density-dependent growth is restricted to juvenile fishes (or small ones such as Norway pout) and the age of maturation does not seem to be affected by changes. Again, density-dependent fecundity is not a sufficient cause for changes in stock-dependent recruitment.

Iles (1973) related the l_1 of the California sardine to year class strength, as noted above, but he was able to reclassify the observations according to the ratio of recruitment, R, to parent stock, P. When R/P is low, the mortality from stock to recruit is high, and vice versa. Three groups of observations were segregated in the ratio R/P: high, medium, or low. The slopes were significantly different from zero; the dependence of growth on numbers sharpens with increasing mortality from parent stock to recruitment. Hence growth and mortality must be linked.

Such observations in the sea have a possible theoretical base. With results on the growth and death of fish larvae (Sette, 1943; Harding and Talbot, 1973; and R. Jones, 1973b), Ware (1975) has established a relationship between the

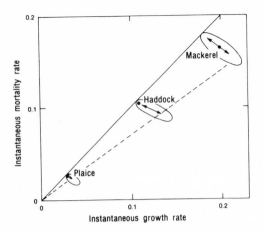

Figure 49. The dependence of the mortality rates of larvae on their growth rates, comparing plaice, haddock, and mackerel. Adapted from Ware, 1975.

growth rates and death rates of fish larvae (Fig. 49). This figure may well be a demonstration of the Ricker-Foerster thesis (Ricker and Foerster, 1948) that when food is abundant fish larvae grow quickly through the predatory fields and suffer less mortality than if food were scarce. Such a mechanism could generate density dependent growth and density dependent mortality, as modeled in a rough way by Cushing and Harris (1973). Ware suggests that

$$M = Q''G', \tag{35}$$

where Q'' is a coefficient, which implies that mortality rate might be estimated from growth. The relationship between growth and mortality could be established for a single cohort if samples of successive ages are not biased by the selections of different gears. Allen (1951) sampled trout in such a way in the Horokiwi stream in New Zealand and plotted numbers on the weights of individuals in the well-known Allen curve: production of the cohort is expressed as function of numbers which decline as it grows older. Neess and Dugdale (1959) analyzed the development of a cohort with this method and showed that

$$- (M/G') \ln (N_t/N_0) = \ln (W_t/W_0). \tag{36}$$

Their logarithmic Allen curve for an insect cohort showed that the ratio $(- M/G')$ was constant throughout the life cycle. Figure 50 shows Mortensen's (1977) data, which suggest that the ratio may remain roughly constant from hatching to the critical size (where specific growth rate equals specific mortality rate).

Ware's generalization ($M = Q''G'$), Le Cren's experiment, and Iles's observations suggest that specific growth rate is inversely related to mortality, or

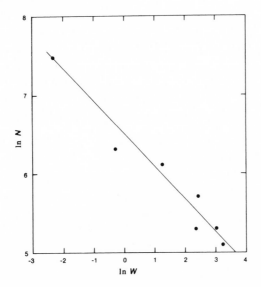

Figure 50. The logarithmic Allen curve (ln N on ln W within a cohort), which suggests a possibly constant ratio of (M/G'). Data from Mortensen, 1977.

to logarithmic numbers. If growth were density dependent, then so would be mortality. Ware proceeds:

$$G' = \delta * W^{-b_1},$$

where $\delta *$ and b_1 are constants.

$$dN/dt = -MN = -Q''\delta * W^{-b_1}N \qquad (37)$$

If δ were inversely proportional to numbers, mortality becomes density dependent because the growth rate depends upon density. Thus the specific growth rate decreases with numbers, and as mortality increases with numbers, so does Q'', but mortality is density dependent only because growth is so.

Summary

Growth has been examined in four ways — in mathematical description, in the comparative study of growth constants in a range of groups, in the experimental ecology of the predator/prey relationship, and in its particular role in the population dynamics of fishes. The growth equations used for different purposes are reviewed, particularly with respect to the transfer of food to weight increment and to the transfer of energy to the water in swimming. The comparison of growth parameters has shown that the ratios M/K and L_m/L_∞ appear to be constant within groups. The first ratio expresses the fact that fish that die young grow quickly, and vice versa; the second ratio describes the character of the reproductive load. The amount of food available to the fish depends upon a variety of factors in the predator/prey relationship. Herring stock changes have occurred, associated with changes in the density of *Calanus*, the preferred food of herring in the sea, perhaps because it is of the optimum size as postulated by Ivlev. Ivlev also showed that the daily ration depended upon the food's concentration and patchiness. Holling's method of separating the predatory processes into components of time will provide further information on the predatory or feeding process.

The role of growth in population dynamics is subtle and complex, more than the mere increment of weight. Any animal eats and grows to avoid death, and the bigger fish are the survivors. Growth is density dependent among juvenile fishes in the sea and perhaps among the larvae. Although the growth of adult fishes does not appear to depend on abundance, the effects of juvenile density dependence might appear among the adults. It is possible that in larval and juvenile life, numbers are reduced until there is more than enough food for the survivors. Then we would expect growth to be density dependent initially and the density dependence to decline until individuals could grow at the best rate permitted metabolically. Hence, growth processes may be at the center of the population regulatory mechanisms.

5
The Measurement of Abundance

In fisheries biology, the basic measure of abundance is the catch per unit of effort, or stock density. If a trawler catches many fish in a given area, there are many fish in that part of the sea; if its catch is small, there are few. Such sampling is used in many studies outside the field of fisheries biology. The essential mathematics of fisheries biology shows how the catch per unit of effort is used as an index of the stock in the sea, and how it reflects the limitations of this index of abundance.

Catch per Unit of Effort as an Index of Stock in Numbers or in Weight

The following treatment is taken from Beverton and Holt (1957). From the growth rates and death rates of the stock, an equation is derived which relates catch (or yield per recruit) to fishing mortality. Fishing mortality is assumed to be proportionally related to fishing intensity (or fishing effort per unit area), so the equation can be adapted to relate catch to fishing intensity. Similarly, catch per unit of fishing mortality (or catch per unit of effort) can be related to stock. The equations will first be derived in respect to numbers in the stock.

Consider the number of fish in the sea, the stock of fish, the natural unit of population. A group of fish joining a stock in a given year survives for a number of years until the animals all die. Such a group is called a cohort. Any stock comprises a number of cohorts, or year classes hatched in succeeding years. The magnitude of the cohort at the age at which it joins the adult stock is called the recruitment, or the year class strength.

If the stock is to remain the same in size, the numbers recruiting to it must equal the numbers dying. Let the number of recruits entering the fishery be R'. After one year,

$$N_1 = R' \exp(-Z), \qquad (38)$$

where N_1 is the number of fish in the year class at the end of one year's fishing, and

Z is the instantaneous coefficient of total mortality.

In the year class, the number dying in the first year is

$$R' - N_1 = R' - R' \exp(-Z) = R' [1 - \exp(-Z)]. \tag{39}$$

The mortality rate, Z, can be separated into two parts:

$$Z = F + M,$$

where F is the instantaneous coefficient of fishing mortality, and

M is the instantaneous coefficient of natural mortality.

The proportion of deaths from fishing to total deaths is given as F/Z.

If the number dying in the first year of the year class is as given in Equation (39), the number caught is

$$(F/Z)R' [1 - \exp(-Z)] = R' E, \tag{40}$$

where E is the rate of exploitation during that year, i.e.,

$$E = (F/Z) [1 - \exp(-Z)].$$

But in subsequent years, the expectation of capture, or exploitation rate, is F/Z. For, after λ years,

$$N_\lambda = R' \exp(-Z\lambda), \tag{41}$$

and the catch after λ years is

$$(F/Z)R' [1 - \exp(-Z\lambda)],$$

where λ is the fishable life span in years between entry to the fishery and extinction.

This expression represents the catch of a year class living for λ years in the fishery; then

$$C = (F/Z)R' [1 - \exp(-Z\lambda)], \tag{42}$$

where C is the yield as catch in numbers.

If λ years are represented in a cohort, there are also λ age groups in any one year in the fishery. If the stock is in a steady state, that is, R', F, and Z are constant, then the annual catch of all age groups is given by the same expression, $(F/Z)R' [1 - \exp(-Z\lambda)]$. This is the case because, in all age groups in one year, the same items are being summed, as in the age groups within a single year class. So the equation $C = (F/Z)R' [1 - \exp(-Z\lambda)]$ also expresses annual catches in a steady-state system. Thus, as $\lambda \to \infty$, $C \to FR'/Z$. It is called the catch equation (due originally to Baranov, 1918), and today it is used considerably in the estimation of stock and/or fishing mortality, as will be described below.

If changes are considered that are due to mortality in the fully recruited adult stock only, it is convenient to work in C/R', the yield per recruit, so that the changes due to natural fluctuations can be ignored. Then,

$$C/R' = (F/Z)[1 - \exp(-Z\lambda)]. \tag{43}$$

This is the equation for yield per recruit in numbers. Similarly,

$$C/F = R'[1 - \exp(-Z\lambda)]/Z. \tag{44}$$

This equation, in yield per unit of fishing mortality, is the theoretical expression of catch per unit of effort, or stock density. The expression $R'[1 - \exp(-Z\lambda)]/Z$ represents the total stock in numbers throughout a year class, or the annual total stock in numbers of all age groups in a steady state.

Fishing mortality is assumed to be proportional to fishing intensity:

$$F = qf, \tag{45}$$

where f is fishing intensity, and

$\quad q$ is a coefficient of proportionality, the catchability coefficient.
So

$$C/F = C/qf, \tag{46}$$

which is catch per unit of fishing intensity. Fishing intensity is fishing effort, g, per unit area, A, so

$$f = g/A. \tag{47}$$

Because $R'[1 - \exp(-Z\lambda)]/Z$ represents total stock and equals

$$CA/qg,$$

which is catch per unit of fishing intensity, then the catch per unit of effort in numbers is an index of stock in numbers. If the catchability coefficient changes, the catch per unit of effort changes its value as an index of stock. In other words, use of the index of stock density must take into account changes in gear, efficiency, and power of ships.

If, however, the catchability coefficient remains constant, the catch per unit of effort, or stock density, is a proper index of stock. Then we may return to the catch equation in numbers to find how the stock density declines as fishing mortality increases. Equation (44) is

$$C/F = R'[1 - \exp(-Z\lambda)]/Z,$$

or

$$C/F = \bar{N}, \tag{48}$$

where $\bar{N}$ is the average stock during the life of the cohort or the average stock in a given year under constant recruitment. Figure 51a shows the decline of

C/F, or stock density, upon F (with q constant and $F = qf$), where C/F is estimated as R' $[1 - \exp(-Z\lambda)]/Z$ for three values of M. The intercept on the ordinate is R' $[1 - \exp(-M\lambda)]/M$. Y/F, or catch per unit effort in weight, also depends on F, in a negatively curvilinear manner, for three stocks — plaice, sprat, and cod.

An important approximation is due to Schaefer (1954, 1957; Schaefer and Beverton, 1963). He developed the logistic equation (see Chapter 7) as follows:

$$dP/dt = a'P(P_m - P) - FP, \qquad (49)$$

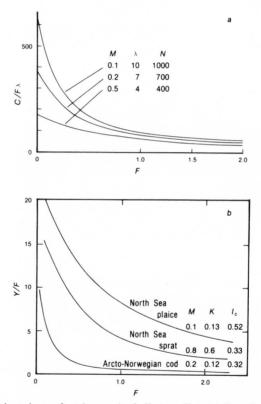

Figure 51. The dependence of catch per unit of effort on effort. (a) From the catch equation in numbers, C/F (stock density) is estimated in numbers and is plotted on F as an index of fishing effort. The intercept in the ordinate is R' $(1 - \exp - M\lambda)/M$; the three curves represent three types of fish with linked values of M (instantaneous coefficient of natural mortality) and R' (number of recruits entering a fishery). (b) Catch per unit of effort in weight, as represented by (Y/F) on fishing mortality, F, estimated for plaice, cod, and sprat (from tables in Beverton and Holt, 1964), for appropriate values of M, K (the rate at which L_∞ is reached, in the von Bertalanffy equation) and l_c, where l_c is the length at first capture.

where P is stock in weight,

P_m is maximum stock in weight,

a' is a constant, the natural rate of increase.

Then

$$FP = a'P(P_m - P) = Y. \tag{50}$$

Catch per unit of effort, Y/f, is a proper index of stock and can be substituted for P.

Then

$$F = a'(P_m - P) \tag{51}$$

and

$$qf = a'[(Y/f)_{max} - (Y/f)], \tag{52}$$

or

$$(Y/f)_{max} - (Y/f) = qf/a'. \tag{53}$$

Thus the decline in catch per unit effort, or stock density, under the pressure of fishing effort is inversely and linearly proportional to fishing effort.

Catch per unit of effort is frequently expressed in weight, kg/100 hr fishing. The weight of a cohort decreases with increased fishing. Beverton and Holt (1957) developed methods by which the changes in numbers and in weight can be combined. They first expanded the growth equation

$$W_t = W_\infty \sum_{n=0}^{3} \Omega_n \exp [-nK(t - t_0)], \tag{54}$$

where Ω_n is the summation constant used in the expansion of the cubic equation and $\Omega_0 = +1$, $\Omega_1 = -3$, $\Omega_2 = +3$, and $\Omega_3 = -1$.

In Equation (41) the change in numbers is described with time, and only one exponent, the total mortality rate, has been used. With weight is added another exponent, K, which does not vary with time in the same way as Z. The two exponents must be combined in such a way that the rate of change of catch can be described with respect to time.

$$dY/dt = FR'W_\infty \exp [-Z (t - t_c)] \sum_{n=0}^{3} \Omega_n \exp [-nK(t - t_0)], \tag{55}$$

where t_c is the age of recruitment to the fishery, or the age at first capture. After grouping the time terms

$$dY/dt = FR'W_\infty \exp (Zt_c) \sum_{n=0}^{3} \Omega_n \exp [nKt_0 - (Z + nK)t] \tag{56}$$

and integrating from t_c to t_λ,

$$Y = FR'W_\infty \exp (Zt_c) \sum_{n=0}^{3} \Omega_n \exp (nKt_0) \int_{t_c}^{t_\lambda} \exp [-(Z+nK)t]dt$$

$$= FR'W_\infty \sum_{n=0}^{3} \{\Omega_n \exp [-nK(t_c - t_0)]/(Z + nK)\} [1 - \exp - (Z + nK)\lambda]. \quad (57)$$

Again, yield per recruit is expressed as Y/R'; catch per effort in weight is expressed as Y/F, or

$$Y/F = R'W_\infty \sum_{n=0}^{3} \{\Omega_n \exp [-\underline{n}K(t_c - t_0)]/(Z + nK)\}[1 - \exp - (Z + nK)\lambda]$$

$$= \overline{P}_w, \quad (58)$$

where $\overline{P}_w$ is the average stock in weight.
In other words, the catch per effort in weight,

$$Y/F = \overline{P}_w, \quad (59)$$

is an index of stock in weight.

Figure 51b shows the dependence of catch per effort in weight upon effort as $\overline{P}_w$ on F. The decrease is less sharp with F than that of catch per effort in numbers (Fig. 51a). The loss in weight with increased fishing mortality is additional to that in numbers. Furthermore,

$$Y/qf = Y/F,$$

and so the estimates of stock density depend upon the catchability coefficient. The theoretical statement is based on the dependence of catch per effort, as index of stock, on abundance, and this dependence is estimated by the catchability coefficient. In reality, "catch per unit of effort" is obviously not the same index of abundance at all times. Fish may be vulnerable to some gears, yet not to others. They may be accessible to fishing fleets at one season, but inaccessible at another. Changes in a ship's power or in gear efficiency may take place. Vulnerability and accessibility are terms describing measurable modifications of the catchability coefficient (see below).

Cohort Analysis: A Development of the Catch Equation

Following R. Jones (1964), Murphy (1965) took the ratio of catch equations in successive years of a cohort. In the first year,

$$C_i = N_i [F_i/(F_i + M)] [1 - \exp (-F_i - M], \quad (60)$$

where i indicates the ith year, and
N_i is the number alive at the beginning of that year.

In the following year,

$$C_{i+1} = N_{i+1} [F_{i+1}/(F_{i+1} + M)] [1 - \exp(-F_{i+1} - M)]$$
$$= [N_i \exp(-F_i - M)][F_{i+1}/(F_{i+1} + M)][1 - \exp(-F_{i+1} - M)]. \quad (61)$$

Then the ratio of catches in successive years is

$$C_{i+1}/C_i = F_{i+1}[1 - \exp(-F_{i+1} - M)] (F_i + M)$$
$$[\exp(-F_i - M)]/[1 - \exp(-F_i - M)] (F_{i+1} + M) F_i. \quad (62)$$

There are three unknowns in the ratio of catches in two years. For each added year in the cohort, a new unknown is added.

Gulland (1977) proceeds, guessing F and M in the oldest age group of the cohort. Let

$$V_n = \sum_{i=n}^{\lambda} C_i, \quad (63)$$

where V_n is "virtual population" at age n where λ is the oldest age in the cohort.

$$S_n = V_{n+1}/V_n, \quad (64)$$

where S_n is an estimate of survival in the nth year.

$$E_n = V_n/N_n, \quad (65)$$

where E is the rate of exploitation.

$$r_n = N_{n+1}/C_n, \quad (66)$$

which is the inverse ratio of catch during year n to the population at the start of the next year, $n + 1$.

$$N_{n+1} = N_n \exp[-(F_n - M)], \quad (67)$$

then

$$r_n = \{(F_n + M) \exp[-(F_n - M)]\}/\{F_n [1 - \exp(-F_n - M)]\}, \quad (68)$$

which can be tabulated as a function of F for any given M; consequently, given r_n, F_n can be read off the table. Further,

$$r_n = N_{n+1}/C_n = (V_{n+1}/E_{n+1})/C_n$$

or

$$r_n = (1/E_{n+1}) [V_{n+1}/(V_{n+1} - V_n)] = (1/E_{n+1}) \cdot [S_n/(1 - S_n)]. \quad (69)$$

Then if we know E_{n+1}, r_n can be estimated. From Equation (68) we may calculate F_n and then E_n. Then we may proceed to r_{n-1}, F_{n-1}, and E_{n-1}, and so from older age to younger back up the cohort.

Given the "terminal" estimates of F and M for the oldest age in the cohort, an array of values of F can be calculated by age and year for as long as catches by age in each year are available. Pope (1971) has also introduced a simplified form of cohort analysis. So long as $F > M$, as in most exploited stocks of demersal fishes, good estimates of fishing mortality are obtained with this method. The original method was called virtual population analysis, but the name cohort analysis includes the original and the simplified method and cannot be confused with the method of virtual populations.

More important for our present purpose, the same method yields estimates of stock by year classes and by age, because $r_n = N_{n+1}/C_n$. With a sufficiently long data series, annual estimates of stock can be made. The system is based on the catch equation, i.e., $C = F\bar{N}$, where $\bar{N}$ is the average population during the year; $N = R'[1 - \exp(-Z)]/Z$, where R' is the population at the beginning of the year. Then $F = C/\bar{N}$, an estimate quite independent of $F = qf$.

Cohort analysis is used in stock assessment wherever data on catch-at-age are available for more years than there are in the life span of the stock. The disadvantage, however, is that stock is assessed at the end of the nth year for a quota in the year $n + 2$, and that stock estimate is biased. Indeed, Pope (1971) has shown that the initial estimate of stock (or F) in year n is not corrected by catches at age until the cohort analysis is run backward to year $n - 3$. The problem of the terminal estimate of fishing mortality (which biases the stock estimate) can only be solved by independent estimates of stock or fishing mortality. Such may be made with egg surveys, tagging experiments, estimates of catch per unit of effort (but see below), and perhaps in the future by acoustic surveys.

The Availability of Fish, or the Variability of Catch per Unit of Effort

Although the catch per unit of effort, or stock density, is theoretically proportional to stock, it varies considerably. Marr (1951) used the word "availability" to describe such variation, which is the degree to which fish are available to capture. It is obvious that if stock density were zero, the stock should theoretically be absent, but less obvious that the fish were merely not available to capture.

Vulnerability. Within a given fishery, weather conditions and the behavior of fish are among the factors influencing the catch per unit of effort. Phases of the moon, for instance, can affect catches. Scofield (1929) found that California sardines were caught by purse seines on moonless nights only; presumably they could see the nets on moonlit nights. Further, purse seiners could not shoot their nets when a strong wind was blowing, and the catch per unit of effort was affected directly by the wind strength. Silliman and Clarke (1945) analyzed these conditions by multiple regression, and their adjusted catch per unit of effort is an estimate unaffected by weather conditions. Differences found over a season were rather unimportant, but within weeks they were quite marked. So the California sardines were vulnerable to the purse seine only on moonless nights when the wind was not blowing too hard.

A more comprehensible form of vulnerability is that ascribed to the selectivity of the gear: the escape of little fish through the cod-end meshes of the trawl, the distribution of fish by size in a drift net or a gill net, and the distribution by size on hooks or of shellfish in pots. Cod in the Vest Fjord are caught by longline, as well as by gill net and purse seine. The purse seine is probably a nonselective gear, but the longline and gill net are certainly not. Figure 52 (Rollefsen, 1953) shows the length distributions of cod caught by the three different gears in the Vest Fjord. Obviously, length distributions of the cod stock are selectively different for catches made with longline and gill net as compared with the purse seine. The larger fish, caught only by the purse seine, are invulnerable to longlines and gill nets. There is a considerable literature on the selectivity of fishing gear because we need to know the distribution of fishing mortality with age. (See Chapter 6. On trawls, see Beverton and Holt, 1957; Beverton and Hodder, 1962; Gulland, 1961; R. Jones, 1961. On drift nets and gill nets, see Holt and Thomas, 1957. On hooks, see Rothschild, 1967.)

Another form of capture depends on the density of fish present. As the meshes of a drift net, or the hooks on a longline, fill up with fish, the chances of capture decrease and so the catch per unit of effort decreases as an index of stock as the density of the fish increases. The effect has been named gear saturation (Gulland, 1955a) or, more generally, gear competition (Rothschild, 1977). As more fish are caught, the remainder in the stock become less subject to the fishing power of the gear, which is gear saturation; gear competition occurs when each additional unit of gear competes with previous units, as for

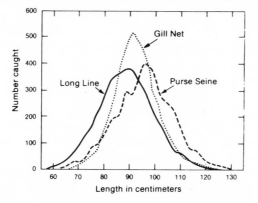

Figure 52. Cod selectively caught by different gears in the Vest Fjord off the Norwegian coast. The length distributions are of fish as caught by longline, gill net, and purse seine. Adapted from Rollefsen, 1953.

example when trawlers aggregate in the Barents Sea, an "arctic city." Rothschild discusses three forms of competition as extension of an assertion that the dependence of F on f is not always a simple linear relationship. The simplest relation is

$$Y/f = (Y/f)_{max} [1 - \exp(-\alpha' f)], \qquad (70)$$

where α' is a coefficient.

The argument was extended by search theory to the complex conditions of purse seine fisheries, where fishermen seek shoals of different sizes of different distances apart under varying weather conditions. If skippers cooperate, as they do to some degree, the probability, p, that one will be successful is

$$p = 1 - (1 - p)^f. \qquad (71)$$

Then an interesting ratio is the proportion of shoals caught cooperatively to those caught independently, h:

$$h = \left(1 - \exp\left\{-f\alpha'[1 - (1 - p)^f]\right\}\right) / [1 - \exp(-\alpha' fp)]. \quad (72)$$

If α', the searching efficiency, is low, cooperation is valuable, but if it is high, cooperation is less and might even become disadvantageous. In such fisheries, catch per unit of effort is not quite a simple function of stock density. Rothschild's third model postulates that fish strike a line of hooks by chance encounter and that saturation occurs when one encounters another caught. There are thus three forms of vulnerability: that affected by the behavior of the fish, that governed by the selectivity of the gear, and that working through the numbers of fish.

Accessibility. Another form of the availability of fish to capture is their accessibility in the vertical or horizontal dimensions. Vertical accessibility can be readily shown in a pelagic species of clupeid, the sprat. Echo surveys conducted in the Wash, an English estuary of the North Sea, have demonstrated that sprat have two distinct layers of vertical distribution — an upper layer of two-year-olds and a lower one of three-year-olds. Figure 53 shows this difference in age expressed in length distributions between the two layers. The sprats are caught by midwater trawls in an area of sandbanks and, because the depth of the trawls cannot be easily controlled, fishermen do not like to tow them too deeply. The histograms show length distributions sampled by a trawl from a research vessel; distribution from the commercial fishery is shown at the bottom of the figure. Hence, the trawls catch only the fish in the top layer, and so only the two-year-old fish are accessible to the fishery (P. O. Johnson, personal communication).

A horizontal form of accessibility is found in the Pacific Ocean. Americans fish for albacore by chumming (where live bait is tossed over the side and

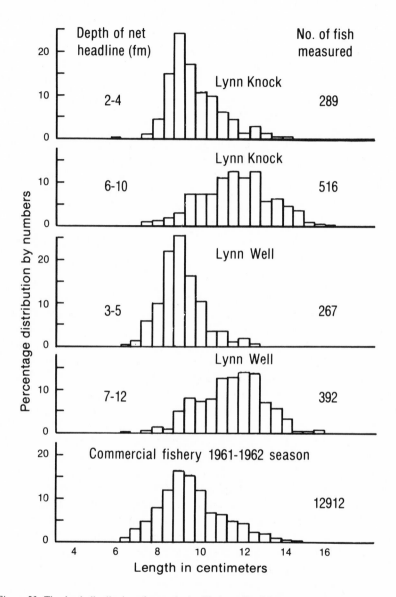

Figure 53. The depth distribution of sprats in the Wash, an English estuary of the North Sea, on 8 January 1962. The fish are vertically distributed in two layers in sandbank areas known as Lynn Knock and Lynn Well. In the top layer at 2–5 fm, the mean length distribution is about 9 cm, and in the bottom layer at 6–12 fm it is about 12 cm. Courtesy P. O. Johnson.

everyone angles among it) and by purse seine off the California coast. The Japanese fishermen exploit what is probably the same stock along the Equator, and again in the Kuroshio (the Black current off Japan) and in the Kuroshio extension, working progressively eastward during the summer (Van Campen, 1960). Each group of fishermen might regard their fishery for albacore as exploitation of an American or Japanese stock. In view of the albacore's life in the North Pacific gyre, within which it makes transpacific migrations, either view is very unlikely. The stock is accessible to one fishery at a time, but it is possible that all the albacore fisheries in the North Pacific should be grouped into one North Pacific stock, living in the North Pacific gyre.

There are two forms of accessibility, vertical and horizontal. Accessibility and vulnerability obviously must overlap on some occasions and are facets of a more general concept. The word "availability" includes the meanings of both words, "accessibility" and "vulnerability." It is a useful term not only because it includes the dichotomy of presence and absence, but also because it expresses the possibility that fish can be accessible and still be to some degree invulnerable. A formal definition of availability emerges in the next section.

The Use of the Catchability Coefficient

The measure of abundance is based on two assumptions that are of funda-mental importance to the population dynamics of fisheries biology. The first assumption has already been noted: $F = qf$ [Equation (45)]. It is a reasonable assumption to make when trawling for bottom-living fish, which are randomly distributed with respect to the fishing gear. It is also true for schooling fish, so long as they are distributed independently of the fishing gear. But if fishermen aggregate on to patches of fish this assumption is no longer valid, as noted above, if the units of gear compete with each other. Even if ships concentrate on areas of fish abundance, with no competition between units of gear, esti-mates of catch per unit of effort must be corrected for the effects of concentra-tion. Many of the seasonal and regional biases in catch per unit of effort as index of abundance were discussed at the Abundance Symposium of the International Council for the Exploration of the Sea (Gulland, 1964).

Catch statistics are collected from an area that has been divided into statisti-cal rectangles. Then the mean density in a heavily fished area can be com-pared with that over the whole area. If each rectangle is fished, it is assumed that within each rectangle the fish are randomly distributed. The concentration of fishing effort as a consequence of the concentration of fish is expressed (Beverton and Holt, 1957) as

$$\sum_{\text{all } i} Y_{ij} / \sum_{\text{all } i} (Y/g)_{ij}, \tag{73}$$

where Y_{ij} is the catch in the ith rectangle in the jth period, and $(Y/g)_{ij}$ is the catch per effort in the ith rectangle and in the jth period. This ratio has been called the effective overall fishing intensity, $\bar{f}$, or concentration factor. The effectiveness of this technique in dealing with shoals of fish and aggregations of ships depends on the size of the rectangle relative to that of the shoal. If the rectangle is so large with respect to the sizes of the shoals that there are many separate aggregations of fishing vessels within it, the method is obviously of little value. If the aggregation of fishing vessels is spread over a large number of rectangles, then a measure of concentration is being expressed in the catches weighted by the distribution of stock density. The East Anglian drift-ermen shot their nets, which were 2 miles long, over an area covered by perhaps twenty or thirty small statistical rectangles (each 9 miles × 9 miles). The catch data by rectangle can be used to describe the aggregation of the fleet on the shoals.

The second assumption concerning catch per unit of effort is that

$$d = q'D, \tag{74}$$

where d is catch per unit of effort in numbers (or estimated density),
$\quad D$ is the true density of the stock, and
$\quad q'$ is a coefficient.
In the terms of Equations (38–42), Gulland (1955a) has shown that

$$C/g = d \text{ and } C/N = F,$$

where N is the number of fish in the stock.

$$\therefore FN = C = gd. \tag{75}$$

Then

$$q = F/f = (C/N)/(g/A) = (C/g)\,(A/N) = d/(N/A)$$
$$= d/D = q'. \tag{76}$$

The two coefficients, q and q', are the same. So the dependence of catch per unit of effort on true density uses the same coefficient as the dependence of fishing mortality on fishing intensity, the catchability coefficient. Formally,

$$D = (C/f)\,(1/q). \tag{77}$$

The deviation of catch per unit of effort from true density is a function of the catchability coefficient. Similarly, changes in availability are well expressed as deviations from the catchability coefficient. Differences in availability from year to year can be expressed in terms of q_0 and q_1, where q_0 is the catchability coefficient in the first year and q_1 is that in the second year. For example, $d_0 = q_0 D_0$ for the first year, and $d_1 = q_1 D_1$ for the second year. It will be shown in Chapter 6 that

$$N_1/N_0 = \exp(-Z) \text{ or } N_0/N_1 = \exp Z. \therefore \ln(N_0/N_1) = Z.$$

Here d_0/d_1 in catches per unit of effort is taken to represent N_0/N_1.

$$Z = \ln (d_0/d_1) \text{ or } \ln (q_0D_0/q_1D_1)$$

So the availability change is given by $\ln (q_0/q_1)$, and the mortality rate is overestimated or underestimated to this degree (Cushing, 1959a). Or the estimate of mortality requires that $q_0/q_1 = 1$ and, theoretically, biases in mortality rate might be detected in an examination of the catchability coefficient.

Dependence of Catchability on Abundance. Another way of estimating catchability is to use independent estimates of stock, as, for example, by egg survey. Stock is then estimated quite independently of catch per unit of effort. Figure 54 shows the relation between the annual egg production of the plaice and the catches per effort in the spawning areas, where the stock is probably well represented. The equation for the lines drawn through the origin in Figure 54 may be expressed as

$$N = (1/q) \, (C/f). \tag{78}$$

The slope is an estimate of the catchability coefficient. It has decreased by three times between the period 1913–21 and 1936–50, and so the catchability coefficient has increased by that amount. In other words, the unit of fishing intensity has become three times as efficient between the two periods. Stock in 1921, after the First World War, was twice that in 1913, and that in 1947 and 1948, after the Second World War, was about double that in 1936 and 1939,

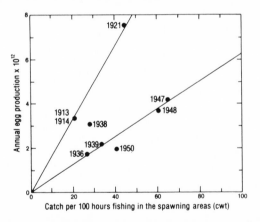

Figure 54. The increase in efficiency of plaice fishing in the southern North Sea. The annual egg production is scaled on the ordinate, and on the abscissa is given the catch per 100 hr fishing in four statistical rectangles over the spawning area. The slope of the line is the reciprocal of the catchability coefficient. Between 1921 and 1936, it appears that the trawlers became three times as efficient in catching plaice. Adapted from Simpson, 1959.

before the war. The catchability coefficient increased by a factor of three between the two wars, and the increase was independent of stock.

Relative efficiency in any one year may be estimated by dividing the effective overall fishing intensity, $\bar{f}$, by the nominal effort. Cushing (1959a) used $\bar{f}$ · [(number of squares)/(number of shots)] in the East Anglian fishery to estimate the increase in efficiency as the driftermen changed from within-square searching to between-square searching. Because the number of squares searched had been reduced since 1936, a correction factor was used and it was shown that the efficiency of driftermen increased during the 1950s. Cushing and Bridger (1966) used another factor, the ratio in each year of catch in numbers per unit of effort to stock in numbers:

$$(C/g)(qg/C) \left\{ [1 - \exp(-Z)]/Z \right\} = q[1 - \exp(-Z)]/Z. \quad (79)$$

The threefold decline in catch/effort corrected by this factor between 1956–60, and 1961–64 was shown to be matched by the same decline in larval abundance (Cushing, 1968a), which implied that q increased with decreased stock.

At the time this phenomenon was thought to be one peculiar to the changing nature of the East Anglian fishery. However, Paloheimo and Dickie (1964) from a model of fisheries on shoaling species suggested that catch per unit of effort remained constant with decreasing abundance as if the number of shoals decreased but shoal density remained constant. But since the use of cohort analysis has made good estimates of stock available, the phenomenon was shown to be somewhat more general. For example, for three distinct cod stocks, Pope and Garrod (1975) showed that the catchability coefficient varied inversely with stock. With material on juvenile herring in the immature fishery for herring of the Atlanto-Scandian stock, Ulltang (1976) showed an inverse relationship between F and numbers in logarithms. McCall (1976) showed that in the California sardine fishery, q was inversely related to biomass. Hence

$$q = \alpha'' N^{-\beta'}. \quad (80)$$

Provided that $(N/\bar{N})$ is constant from year to year,

$$C/f = \alpha'' N^{(1-\beta')}. \quad (81)$$

If $\beta' = 1$, catch per unit of effort is independent of stock; if $\beta' = 0$, catch per unit of effort is proportional to stock. Figure 55 shows the inverse dependence of q on stock for the Norwegian immature herring fishery which was exploited by purse seines; in this example $\beta' > 1$, which may be biased upward.

Fish concentrate in space and time. The seasonal variation in catch per effort may amount to a factor of one or two or of an order of magnitude or so with the highest values at the season of spawning. There is only a difference

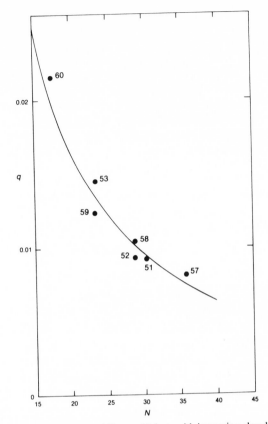

Figure 55. The decline of *q*, the catchability coefficient, with increasing abundance (in thousand millions) in the purse seine fishery for the Atlanto-Scandian herring. Adapted from Ulltang, 1976.

of degree between such variation in stock density and the more extreme one characteristic of shoaling species such as sardine, herring, or mackerel. Hence it is likely that $0 < \beta' < 1$ and that values for flatfish are low, for herring are high, and for gadoids are intermediate. The implication is that catch per unit of effort, or stock density, is not the good index of stock as suggested above. However, it is possible that β' estimates the degree of patchiness or the seasonal variation in stock density; in other words, it may be characteristic of a given species or group of species. If that were the case, catch per unit of effort or stock density would remain a good estimate of stock, independent of the estimates made by cohort analysis.

Catch per unit of effort, or stock density, is no longer the simple index of abundance it was once thought to be. If we suppose that catchability varies inversely with abundance wherever fishermen aggregate onto patches of fish, we should believe it to be a general phenomenon. Then we should examine how such material could be managed.

Some Validations of Catchability. The salmon fisheries on the west coast of North America provide a unique way of testing the validity of the catch per unit of effort as an index of stock density. As the fish come in from the ocean to spawn, they are subject to various fisheries in the estuaries. After passing through a fishery, the salmon move upstream to spawn. They are counted by various means, the most common being visual counts at weirs, and the estimate of total population obtained in this way is called the escapement, i.e., from the fishery. The sum of escapement and catch is called the total run, an estimate of stock. Figure 56 shows the relationship between escapement and catch per unit of effort and that between total run and catch per unit of effort in the fishery for sockeye salmon in Bristol Bay, off the Alaskan coast in the Bering Sea (Tanaka, 1962). In this instance, the catch per unit of effort is a good index of abundance — one that is closely proportional to the total run throughout its range. But the relationship between total run and catch is a poor one, as might be expected over a long period, as the fishery takes different proportions of the stock from year to year.

Herring lay their eggs on the bottom. On the Sandettié spawning grounds, they cannot be counted readily because the stones on which the herring spawn are too big for the grab. However, herring larvae can be caught and counted,

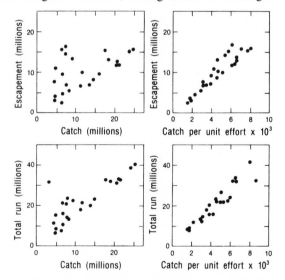

Figure 56. Relationship (1) between escapement and catch, (2) between escapement and catch per effort, (3) between total run and catch, and (4) between total run and catch per effort in the salmon fishery in Bristol Bay on the Alaskan coast of the Bering Sea. Adapted from Tanaka, 1962.

and a relationship established between catch per unit of effort and larval abundance; Bridger (1960) showed that the stock density was corrected for efficiency changes as described above. The slope of the line is not the reciprocal of the catchability coefficient, but is a function of it because the mortality from egg to larva is unknown. However, differences in catch per unit of effort are closely correlated with differences in larval abundance. This is an important conclusion, suggesting that the shoals of fish are randomly distributed with respect to the gear. Fish that shoal, despite their highly nonrandom distribution, can still be distributed randomly with respect to the gear. Hence the corrected catches per unit of effort in the East Anglian herring fishery were probably good indices of stock density. It remains possible to establish the dependence of catch per unit of effort on stock where the dependence of the catchability coefficient on abundance is low and, further, to correct it where it is high.

In recent years a quite different approach has been started by Harden Jones and his colleagues (Harden Jones et al., 1977) with the Admiralty Research Laboratory scanner, and beautiful pictures of the seabed are displayed with a single pulse. Plaice were tagged with transponding tags so that they could be observed during the act of capture. In this way, the efficiency of the Granton trawl in catching plaice could be estimated — about 80 percent with ticklers and 40 percent without them.

It will be recalled that fishing intensity is fishing effort per unit area (g/A, or hours fishing per unit area). Then the catchability coefficient may be defined as *fishing mortality per unit fishing intensity* or *fishing mortality per hours fishing per unit area*. The coefficient is usually expressed in units of time because the area swept by the trawl has been considered inaccessible. The results of the efficiency experiments suggest that in the example described above the catchability coefficient could be estimated as *fishing mortality per hour's fishing per (0.8) area swept*.

Another consequence of this work on trawl efficiency is that distributions of stock density of plaice [i.e., catches per 100 hr fishing per (0.8) area swept] could be used theoretically to make estimates of stock, as in a groundfish survey. Then catches on the same scale could be used to estimate fishing mortality directly. At the present time this method of estimating trawl efficiency is in its infancy and the theoretical extensions have not been yet tried.

The Development of Acoustic Methods to Estimate Stock

The recording echo sounder was developed for the navigational purpose of making a quasi-continuous record of the depth of the sea. From a nickel or ceramic transducer in the ship's hull an ultrasonic pulse is transmitted into the sea at the speed of sound, about 1500 m/sec. The pulse is reflected from the seabed to the transducer by the shortest path, which is the depth. On the paper record the transmission signal is indicated and the depth in m is given by half the time interval in seconds multiplied by the speed of sound in m/sec, e.g., (0.133 sec × 1500 m/sec)/2 = 100 m. Signals from fish from midwater are indicated between those of transmission near the surface and those of the seabed. The first records date from the 1930s, including a famous one of cod in the Vest Fjord (Sund, 1935).

The recording echo sounder yields many rolls of paper as the ship steams across the sea. An obvious step to the fisheries biologists just after the Second World War was to make charts of the horizontal distribution of the records of fish. Provided that the fish remain at about the same depth and the weather is not too rough, this method remains a useful way of making estimates of relative abundance. However, the records must be identified as the desired species by capture, because there is no method of acoustic identification. Fish that appear, for example, in 100 m depth in March in the inlet of Port Susan, in Puget Sound, are almost certainly hake, but such identification is made by association with place and season.

The first step toward a quantitative estimate of abundance was made by recognizing the signal from a single fish as a received signal of one pulse length for each transmission as the target crossed the beam. Midttun and Saetersdal (1957) carried out surveys of single fish, in about 150 m in the Barents Sea, which were almost certainly large cod. As a consequence it became necessary to calibrate transducers and to make estimates of the signal to be expected from a single fish. The "target strength" of a single fish increases as about $W^{0.72}$ and increases somewhat at higher frequency. Goddard and Welsby (1977) made very large numbers of measurements on caged live gadoids at three frequencies. Nakken and Olsen (1977) made measurements on stunned or freshly killed fish related in the echo-sounder beam in the fore-and-aft dimension at two frequencies. Their results are given in Figure 57, relating target strength to log length. Units are in decibels (dB), which are logarithmic ratios (× 10) of intensity or pressure; because the decibel is a ratio it must be related to a standard, in this case a unit of area, a sphere of 2 m radius (for target strength, T''). The figure expresses the dependence of scattering cross-section (or target strength) on length, both parameters being normalized for frequency. The variance about the regression is high, about an order of magnitude; the regression is probably well estimated, but the variance might have been imposed by the experimental method.

The second step toward estimating abundance was made in estimating the volume from which the signal from a single fish of a given size might be received. An acoustic survey of hake was carried out between Cape Town, in South Africa, and Walvis Bay, in Namibia. The fish were recorded as single targets within one fath of the seabed. The echo sounder is a noise-limited instrument, which means that any signal is received relative to a background noise from the water, the ship's propeller, and the slap of waves on the hull.

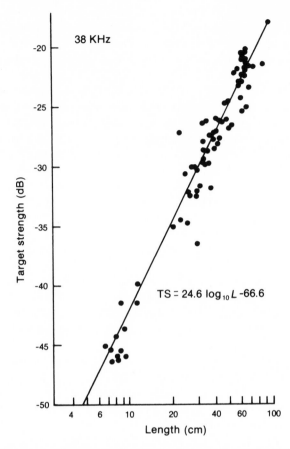

Figure 57. Dependence of target strength on log length of cod. Target strength is an estimate of acoustic cross-section of a target, in this case a fish. It is a power function of length. Adapted from Nakken and Olsen, 1977.

For a fish of a given size, there is a maximum range at any angle to the transducer defined by a conventional signal-to-noise ratio; the directivity pattern of the transducer was known, and so the volume in which the fish lived was calculated. Figure 58 shows the density of hake estimated with this method (Cushing, 1968b).

The calculations are based on the use of a sonar equation, which for a single fish may be written:

$$I = I_0 \ (\sigma/4\pi) \ b^{*2} \ (\theta', \ \phi'), \tag{82}$$

where I is the echo level or intensity received from a single fish,

I_0 is the source level or transmitted acoustic intensity in watts per m^2,

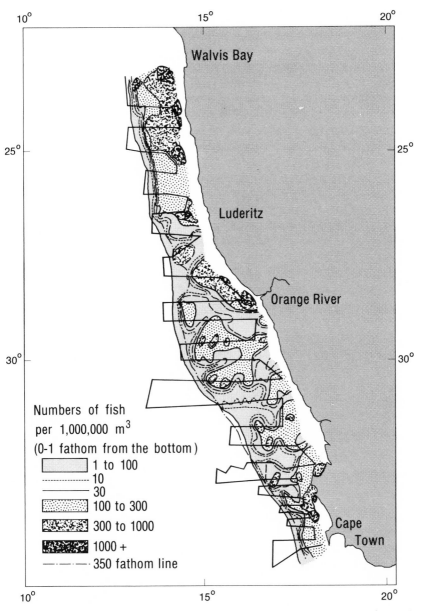

Figure 58. An acoustic survey for hake in one fathom off the seabed off South Africa and Namibia. Adapted from Cushing, 1968b.

$b*$ is the directivity function in the two dimensions of the beam angle, θ' and ϕ', and

σ is the scattering cross-section of the fish in cm^2.

In recent years electronic engineers have devised a time varied gain (TVG), or amplifier, which corrects the received signal for the effects of losses due to the spreading of energy with range (by the inverse square law) and attenuation (due mainly to molecules of magnesium sulphate in the sea). This equation may also be written in the logarithmic form in decibels:

$$E' = S' + T'' + 20 \log b' \, (\theta', \phi'),$$

where S' is the source level in dB re 1 μV,

E' is the echo level in dB, and

T'' is the target strength in dB.

The third step toward an estimate of abundance was made by Bodholt (1969), who introduced another form of sonar equation for use with an integrator which sums all signals for a nautical mile within chosen ranges:

$$I = I_0 \, (\sigma N/4\pi) \, (c'\tau'/2) \, \Omega_0, \qquad (83)$$

where N is the number of fish,

c' is the speed of sound in sea water in m/sec,

τ' is one pulse length, and

Ω_0 is a solid angle, which should be defined by the noise level, and which is inversely proportional to the signal from the scatterers.

Because scattering cross-section, σ, varies as $W^{0.72}$, this method estimates biomass.

The purposes of the two acoustic estimates of abundance are distinct: to estimate the numbers of single fishes (and perhaps their sizes), and to estimate biomass. The traditional stock estimates of fisheries biologists are made in biomass, but it is usually the biomass of fish larger than a minimum length and very often it is a quantity which is conveniently separated by age groups. Biomass, as sampled by an echo sounder, may consist of a number of fish of a given size, but the transducer cannot distinguish this signal from that of a much larger number of much smaller animals because they may be of the same amplitude. If the fish could be counted, this bias is eliminated because signals from single fish of less than a certain amplitude may be disregarded. If the same fish could be sized, the product of numbers and sizes by weight groups is the same biomass used by fisheries biologists. However, shoals of fish cannot be counted; the ideal system is one which counts single fishes and integrates the signals from shoals. It is assumed that signals processed in this way are those received from those fishes caught in the sampling gear. Bias due to the biomass of small animals is reduced, but not eliminated, and the biomass processed in the sampling system is that needed by the fisheries scientists.

Acoustic surveys have been carried out throughout the world for many years. The simplest form of survey for relative abundance is used in fisheries research vessels as part of the normal routines of finding fish. The biomass method with an integrator is used extensively for exploratory work on Pacific hake and southeast Alaskan herring (Thorne, 1977), sockeye salmon (Mathisen, Croker, and Nunnallee, 1977), juvenile fish in the Barents Sea (Haug and Nakken, 1977), anchovy and horse mackerel in the Sea of Marmara, and of pelagic fishes of various species off West Africa (Johannesson and Losse, 1977). Stock estimates of blue whiting (*Micromesistius poutassou* [Risso]) (Midttun and Nakken, 1977; Pawson, Forbes, and Richards, 1975) and of mackerel have been made in British waters with this method. However, more sophisticated acoustic methods as independent checks of the ordinary methods of fisheries biologists are not yet available.

Summary

Starting with stock in numbers or in weight, catch in numbers or yield in weight can be related to fishing mortality. The equations are called the catch or yield equations. From the catch equation, fishing mortality is the ratio of catch in numbers to the average stock in the sea. But fishing mortality is also proportional to fishing intensity by the catchability coefficient and hence, theoretically, catch per unit of effort is a proper index of stock. One would therefore expect catch per unit of effort, or stock density, to be related inversely to fishing effort either in the theoretically curvilinear way or in Schaefer's linear approximation.

An important development of the catch equation in numbers is cohort analysis, by which a series of catch equations by age within a cohort are solved. With information on catch-at-age in numbers for many years, the result is a matrix of fishing mortalities (or numbers in stock) by age and year, but estimates for the three most recent years in the matrix are biased. For quota assessment it is necessary to make use of information on effort or catch per unit of effort.

The catchability coefficient can, however, be affected by behavior selectivity and by gear saturation. Such differences are called changes in availability, whether fish are accessible to the gear or vulnerable to it. But by far the most important bias is the increase in catchability with declining abundance; if we were to rely on catch per unit of effort, or stock density, alone, any decline in abundance would be underestimated. Stock density can be related successfully to numbers estimated by cohort analysis, but independent estimates of stock are needed from time to time by egg survey or acoustic survey. Acoustic surveys are used considerably for exploratory purposes, but in the future they may well be used for population analysis.

6

Estimates of Natural and Fishing Mortality

Fishing is the livelihood of fishermen. It is also of scientific value to fisheries biologists. Fishing kills fish, and both the catch and the number of ships and the time spent fishing can be readily recorded. Therefore, in principle, fishing mortality should be easy to measure because variations in catch per effort can be related to variations in fishing effort, the total time spent fishing by the fleets. Accordingly, a good fisheries biologist starts his studies by attempting to measure the effect of fishing mortality.

This chapter examines briefly the estimation of total mortality, Z, but it is mainly concerned with the separation of total mortality into its two components, fishing and natural mortality. If fishing represents a large fraction of the total mortality, the separation should be made readily. If fishing represents only a small fraction of the total mortality, separation of the two components is rather difficult because the observations are often variable.

Estimation of Total Mortality

The sampling system on the quays gives numbers and weights of fish, their ages, the positions and sizes of the ships, and the time spent fishing. Age distributions are then constructed in numbers of each age group per unit of effort, e.g., so many six-year-old cod per 100 hr fishing (of a standard vessel). Such age distributions are made by weeks, months, or quarters, and are averaged for the year. Provided that the biases in availability are detected and eliminated (particularly the dependence of catchability on abundance, discussed in the last chapter), such estimates are good indices of stock. Then the decrease in stock density (as numbers per unit of effort) of a year class from year to year is a good measure of total mortality. Within a year class, $dN/dt = -Zt$, $N_2/N_1 = \exp(-Z) = S$, or survival index, and

$$N_1/N_2 = \exp Z, \tag{84}$$

where N_1 is the catch per unit of effort in numbers of an age group at the start of a given year, and

N_2 is the catch per unit of effort in numbers of the succeeding age group of the same year class at the start of the subsequent year.

$$\therefore Z = \ln (N_1/N_2). \tag{85}$$

Total mortality may also be estimated from two stock densities of a year class separated by a number of years (for example, when $t = 3$).

$$Z = (1/t) \ln (N_1/N_4)$$

As will be shown below, it is sometimes useful to estimate survival directly in tagging experiments. Independent measures of survival may be used from the stock densities of all age groups in successive years, but because those at older ages are necessarily low, some weighting in age is needed. Chapman and Robson (1960) showed that the best such estimate of *average* survival is the maximum likelihood expression.

$$S = T^* [(\Sigma N) + T^* - 1], \tag{86}$$

where $T^* = N_1 + 2N_2 + 3N_3 + \ldots$, and
$\Sigma N = N_0 + N_1 + N_2 + \ldots$,
which is a weighted estimate of average survival from year class to year class from one year to the next. For most purposes the best estimate is that of the instantaneous coefficient of total mortality within a year class, i.e., Z.

The Relation between Fishing Intensity and Total Mortality

The average abundance of a year class in a given year (estimated usually in stock density), the first of two consecutive years, is

$$\bar{N}_1 = N_1 [1 - \exp (-Z_1)]/Z_1, \tag{87}$$

where N_1 is the stock density of fish at the start of the year,

$\bar{N}_1$ is the average stock density in the first year, and

Z_1 is the total mortality coefficient during the first year.

In the second of the two years, the average abundance of the same year class, $\bar{N}_2$, is

$$\bar{N}_2 = [N_1 \exp (-Z_1)] [1 - \exp (- Z_2)]/Z_2, \tag{88}$$

where $\bar{N}_2$ is the average stock density in the following year, and

Z_2 is the total mortality coefficient during the second year.

Since $Z = \ln (N_1/N_2)$, as in Equation (85), then the average stock density in the first year can be divided by that in the second, as

$$Z = \ln (\bar{N}_1/\bar{N}_2) + \ln \left((Z_1/Z_2) \{[1 - \exp (- Z_2)]/[1 - \exp (- Z_1)]\}\right). \tag{89}$$

The second term in the above equation is the correction term for changes in total mortality from year to year (in this case, from the first year to the second year). So, if Z is constant from year to year,

$$Z = \ln (\bar{N}_1/\bar{N}_2) = F + M, \tag{90}$$

where F is the instantaneous coefficient of fishing mortality, and
M is the instantaneous coefficient of natural mortality.
Since a previously given axiom states that $F = qf$,

$$\ln \bar{N}_1/\bar{N}_2 = qf + M. \tag{91}$$

The fishing intensity, f, has been taken as that exerted in the first of two consecutive years, because the stock abundance in the second year has suffered fishing and natural mortality during the first year. Beverton and Holt (1957) have applied this method to the fishery for the Fraser River sockeye salmon in British Columbia (Fig. 59). Total mortality (or $\ln (\bar{N}_1/\bar{N}_2)$) has been plotted on effort, g, in gill-net units. There is only one age group, so the second term in Equation (89) was not used. The slope of the regression in Figure 59 is an estimate of the catchability coefficient, q; $q = 0.311$, or 1 gill net generates $0.0012\,F$ (from $F = qf$). The intercept of the regression is an estimate of natural mortality, M; $M = 0.648$. Such an estimate based on the intercept is of course variable. In this analysis, the estimate of M includes fishing mortality outside the mouth of the Fraser River, but the catchability coefficient remains valid for the fishery in the river itself. Paloheimo (1961) has shown that if the average abundances, $\bar{N}_1$ and $\bar{N}_2$ and the measures of fishing intensity, are estimated at the midpoint of the year, the correction term is no longer necessary. This is because, when measured from the midpoint of

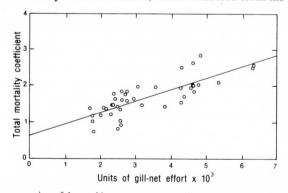

Figure 59. The regression of the total instantaneous mortality coefficient on units of gill-net effort for Fraser River sockeye salmon in British Columbia. The intersect should estimate natural mortality, but in this case it also includes the effect of a fishery in the sea at the mouth of the Fraser River. The slope of the regression is q, the catchability coefficient, or 1 gill net generates $0.0012\,F$. Adapted from Beverton and Holt, 1957.

the first two years to the midpoint of the second, ln $(\bar{N}_1/\bar{N}_2)$ contains equivalent elements of both Z_1 and Z_2 and of f_1 and f_2. Today this is the general way in which this method should be used.

This promising method has not been very successful, however, because the regressions are often variable. Let us recall that $d/D = q$; then $Z = \ln (q_0 D_0/q_1 D_1)$, where D_0 is the true density in the first year and D_1 that in the second, and q_0 is the catchability coefficient in the first year and q_1 that in the second. Hence in any pair of years, Z is overestimated or underestimated by ln (q_0/q_1). Any such regression may conceal a potential bias in that when Z is low, abundance may be high, and vice versa; then catchability may be high when Z is high and vice versa. Although we would not expect (q_0/q_1), the single year's increment in catchability, to be very high, it remains a potential source of bias. Natural mortality is assumed to be constant, whereas it may well increase with age in a lightly exploited stock at low levels of fishing intensity. In a single species fishery, effort can be properly recorded, but in a mixed fishery the effort cannot be readily allocated by the species caught, in the sense that the same effort may generate different fishing mortalities on different stocks in the same catch.

Yet the method works well in carefully selected fisheries. Figure 60 shows the dependence of total mortality summed in year classes on effort in a gill net fishery for *Tilapia* in Victoria Nyanza (Garrod, 1963). It is perhaps successful because the effort is in a sense selected by the fish as they swim into the gill nets. Burd and Bracken (1965) estimated the natural mortality of Celtic Sea herring with this method possibly because the fishery was not mixed at all, i.e., only herring were caught. For the Icelandic cod fishery, Gulland (1972) derived a good estimate of natural mortality by grouping the observations in five-year periods. On the other hand he showed for the arctic cod fishery in the Barents Sea that an estimate derived from six-year-olds

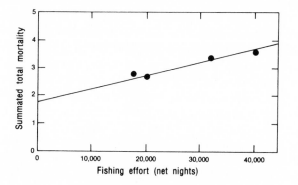

Figure 60. Regression of total mortality on fishing effort in the *Tilapia* gill-net fishery in Victoria Nyanza. Adapted from Garrod, 1963.

differed from that derived from seven-year-olds, yet the exploitation rate on the two groups cannot differ much. Clayden's (1972) estimate on the same stock is much more variable, perhaps because the English fishery exploits the immature stock from which the older fish may emigrate to regions where the adults live. Halliday (1971) estimated the natural mortality of haddock off Nova Scotia with a good regression, perhaps because total mortality was estimated from catch curves, which might have reduced the variability due to ln (q_0/q_1). Each successful use of this method may depend on the judicious use of data to avoid the potential biases, but in a mixed fishery with substantial differences in catchability the method may fail. There are, however, two distinct requirements: first, a good estimate of natural mortality, which is difficult, and second, an estimate of the exploitation rate (F/Z), which is not quite so difficult but which is the essential requirement of any assessment.

Since cohort analysis was introduced by Gulland (1965) and Garrod (1967), estimates of P_n, stock in numbers, and of F emerge for each age and year; the method demands an estimate of natural mortality. The values of F input in the last year of the cohorts are necessarily guessed, but they become corrected as catches at age are added in the calculation as it proceeds from older to younger age groups in the cohort. As noted in the last chapter, Pope (1972) has shown that the bias in the input guess is removed in three years of back calculation. However, Agger, Boetius, and Lassen (1973) demonstrated that if M is overestimated, F will be underestimated, and vice versa; indeed, a biased value of F will not converge, and the bias enlarges with decreasing age as the calculation proceeds. Hence F will be unbiased only if M is well estimated, or nearly so, and there is no way at present of estimating natural mortality well.

Schumacher (1971) examined the West Greenland cod fishery and, assuming that $M = 0.2$, emerged with a highly significant regression of fishing mortality on fishing effort with a zero intercept, from which he concluded that natural mortality was correctly estimated. However, this conclusion also demands that fishing effort be properly described, which is reasonable in such an unmixed fishery, and also that the catchability coefficient is not density dependent (see Chapter 5). Winters and Hodder (1975) used the same argument in a study of the herring of the St. Lawrence River, which might be reasonable because the fishery is an unmixed one. A variant of this method was used by Pinhorn (1975) on the cod stocks on the Grand Banks, off Newfoundland; natural mortality was estimated $(M = 0.2)$ with a regression of total mortality (from catch curves) on effort. The same result emerged when an estimate of *total* mortality from cohort analysis was used. The fishery is predominantly an unmixed one and, as it is an old one, the assumption that natural mortality is constant with age (and exploitation) is not unreasonable. That the method has limitations emerges from Pitt's (1973) study of the American plaice, in which estimates of natural mortality were derived from unexploited stocks and from regressions of total mortality on effort. Yet the regression of fishing mortality (from cohort analysis) on effort yielded a positive intercept, perhaps because natural mortality was higher at low fishing intensity when the stock was virtually unexploited.

This brief review of Beverton and Holt's method of separating fishing mortality from natural mortality reveals that, although sometimes successful, it is not, nor can it be, necessarily so. Results may be biased by differences in natural mortality with age or exploitation, by differences in catchability from year to year due to changes in availability, or by differences in catchability from year to year with stock size. However, the real source of difficulty is that in a mixed fishery effort cannot yet be allocated properly by the species in the mixture. A hundred hours of fishing inflicts different fishing mortalities on each species in the catches, but it is not separated into their components. The method works well under special circumstances, can give useful approximations of exploitation ratio, but needs considerable development before it can add to our knowledge of natural mortality.

Estimation of Fishing Mortality by Tagging

Direct Estimates by Single or Multiple Releases. Tagging experiments made at sea differ in scale from those carried out in lakes and rivers. At sea, tags are released in a large population in which they should eventually mix randomly during its seasonal migrations. Biologists then look for recoveries from year to year in particular fisheries. Hence the experiments are usually based on a single release with the object of estimating survival rate or fishing mortality rate. Fisheries in freshwater can be studied in this way, but the populations are small, and so tagged fish are assumed to mix in the population quite quickly. Hence multiple releases within a season can be conducted with considerable statistical elaboration, if that assumption is justified. Such methods have been used to estimate stocks in numerous and disparate bodies of freshwater. Robson and Régier (1964) published charts which show the sampling error for numbers tagged and different magnitudes of population. However, Gulland (1963) has pointed out that there is a dilemma in the use of data from tagging experiments. Because some fish die from the very act of tagging, the best estimates of death rate in a time series of recaptures would be those made closest to the time of tagging. Because of the mixing of tagged fish with untagged ones, the best estimates are those later in the time series when mixing is complete. Because of the latter bias, the estimated fishing mortality is not really that of the stock. Such effects might be expected at sea where fish may quickly gather from broad regions and disperse as rapidly into the ocean, but, less obviously, the same must be true of lakes, even little ones. Gulland has suggested that the use of tag returns per unit of effort, in the local area around the tagging point, might express results in abundance. Estimates can be made of fishing mortality per unit of fishing intensity which is q, the catchability coefficient. Let

$$n_i/f_i = q\bar{N}_i, \tag{92}$$

where f_i is the fishing effort in area i,
$\quad n_i$ is the number of returns in area i, and
$\quad \bar{N}_i$ is the mean number of tagged fish in the stock in area i.

In other words, the number of returns per unit of effort is proportional to the average abundance of tagged fish in the area. Figure 61 shows the percentage, in logs, of the initial number of plaice tagged that were recovered per 100 days fishing per statistical rectangle in the southern North Sea during the period immediately after the Second World War. The decline in this percentage with time is due to mortality and other losses. The intercept on the ordinate estimates q, the catchability coefficient, 6 percent per 100 days fishing per rectangle. (The intercept is the nominal catchability, (n_i/f_i) $(100/N_0)$, equal to actual catchability, if all the fish survive.) Gulland averaged the estimates of the catchability coefficient for four rectangles, weighted them by the effective overall fishing intensity, and derived the following values of F (from $F = qf$):

$$F = 0.53 \ (1946)$$

$$F = 0.63 \ (1947)$$

$$F = 0.71 \ (1948).$$

In the period before the Second World War, the estimate of fishing mortality for plaice in the southern North Sea was given as $F = 0.73$ (Beverton and Holt, 1957). The virtue of Gulland's method lies in the direct estimation of q,

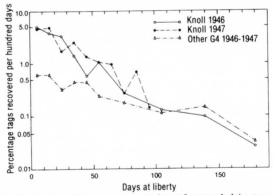

Figure 61. The decline in time of the percentage numbers of returned plaice tags per 100 days for each unit of catch per effort. The intersect is an estimate of q, the catchability coefficient; see text for derivation. Smith's Knoll is a fishing ground in the southern North Sea; G4 is a statistical rectangle in the same region. Adapted from Gulland, 1963.

the catchability coefficient, from the results of the tagging experiment. Here an estimate of fishing mortality becomes possible from measures of fishing effort.

There are two forms of tagging experiments used to estimate fishing mortality, that of a single release and that of a multiple one. In either, the recaptures may be expressed as proportion of the stock in the sea or to the fishing effort exerted on that stock as indicated by Gulland's analysis. By a historical accident, the theory of single release is a deterministic one, whereas that of the multiple release is today a stochastic one. Because the latter yields the most information from a simple experimental arrangement, it may eventually replace the single release.

Beverton and Holt's (1957) treatment considers a population, or substock of tagged fish. Hence their equations are very similar to those used in Chapter 5. The coefficient of natural mortality, M, is extended to an "other loss" coefficient, X, which includes not only the added mortality of tagged fish, but also tag-shedding and the failure to report tags (see below). So the rate of change of the number of tagged fish in the substock of tagged fish is

$$dN/dt = -(F + X)N, \tag{93}$$

where N is the number of fish in the tagged population, at any time, t, and
X is the "other loss" coefficient.

$$N_t = N_m \exp [-(F + X)t], \tag{94}$$

where N_t is the number of tagged fish in the sea after time t, i.e., $(t_1 - t_0)$, and
N_m is the number tagged at the beginning of the time period $(t_1 - t_0)$.
Let $(F/F + X)$ be the proportion of loss by fishing to the total loss in the tagged population. Between t_0 and t_1, the number recaptured, n_1, is

$$n_1 = [FN_m/(F + X)] \{1 - \exp[- (F + X) (t_1 - t_0)]\}. \tag{95}$$

Between t_1 and t_2, the number recaptured, n_2, is expressed by

$$n_2 = \{FN_m \exp [- (F + X) (t_1 - t_0)]/(F + X)\}$$
$$\{1 - \exp [- (F + X) (t_2 - t_1)]\}.$$

If $t_1 - t_0 = t_2 - t_1 = \tau$, i.e., all time periods are equal, then Equation (95) can be divided into Equation (94):

$$(n_2/n_1) = \exp [- (F + X) \tau],$$

$$\therefore (F + X) = (1/\tau) \ln (n_1/n_2). \tag{96}$$

If substitutions are made in Equation (95), then

$$F = [(n_1/\tau) \ln (n_1/n_2)]/[N_m (1 - n_2/n_1)]. \tag{97}$$

Thus, the fishing mortality coefficient is separated from the losses of tags due to all other factors.

The argument may be extended by estimating a regression of $\ln (n_{t+1}/n_t)$ on equal time intervals, where the slope is $-(F + X) \tau$ and the intercept I'_0 (as indeed Beverton and Holt pointed out). R. Jones (1956) wrote:

$$\ln n_i = \ln N_m + \ln F$$
$$+ \ln \left(\{1 - \exp [- (F + X)\tau]\}/(F + X) \right) - (F + X)t. \qquad (98)$$

By rearrangement,

$$F = (F + X) \exp [I'_0 - (F + X) \tau]/N_m\{1 - \exp [-(F + X) \tau]\}. \quad (99)$$

Paulik (1963) plotted $\ln (n_i/N_m)$ on equal time intervals and weighted each value by (n_i/N_m). Then, with an analogous argument,

$$F = (F + X) \exp [I'_0 - (F + X) \tau]/\{1 - \exp [- (F + X) \tau]\}, \quad (100)$$

where I'_0 is the intercept.

If $(F + X) \tau$ is small, Gulland (1969) has shown that the regression may be simplified by plotting $\ln n_i$ on $0.5t$, $1.5t$, etc; the slope is $- (F + X) \tau$, and the intercept is $I'_0 = \ln (FN_m\tau)$. Then,

$$\hat{F} = \exp (I'_0/N_m\tau). \qquad (101)$$

The virtue of using such regressions is that with the assumption of constant survival, the numerous uncertainties of a tagging experiment with a single release in a lake or in the sea are distributed. There are three types of error in such an experiment (Ricker, 1948). The first, type A, includes the loss of tags due to immediate mortality and the added loss due to failures of fishermen to report recaptures; it is expressed by the intercept of the regression of the logarithms of numbers at recaptures on time. The second, type B, includes the mortality of fish due to tagging during the period of the experiment. This form of mortality excludes immediate death in the tagging tank, but includes death due to tagging after release; such errors are expressed in the slope of the regression. The third, type C, is due to emigration from the area of tagging, which would bias the slope of the regression downward; as indicated in Gulland's experiment, the problem is less one of emigration than one of adequate mixture with the real population.

A tagging experiment with herring in the southern North Sea was carried out by Ancellin and Nédelèc (1959) in the winter of 1957–58 on the spawning grounds of the Downs stock of the North Sea herring, where many trawlers work. There are three main trawling grounds (which are also the spawning or assembly grounds shown in Figure 6): Sandettié and Cap Blanc Nez in the

Straits of Dover; Vergoyer off Boulogne; and Ailly off Dieppe. The tagging was carried out on these three grounds during the period of the fishery. There is a migration by the fish from Sandettié westward, from Vergoyer westward, and from Ailly coastward. However, most tags were obtained from the ground of liberation (where the recapture average was 77 percent for all grounds). Figure 62 shows the regular decline in number of recovered tags (in logs) with time on each ground. At Sandettié, the numbers recaptured after 10 days were very few, but after 24 days tagged fish were still being recaptured on the Vergoyer ground. The Sandettié Bank is also a spawning ground, and it is likely that much of the decline might be due as much to emigration after spawning as to any other cause.

The fishery on each ground, for example, the Sandettié ground, lasted for 10 days. Because fishermen moved from one ground to another, the whole fishery on all four grounds lasted longer — 24 days or more. On the basis of

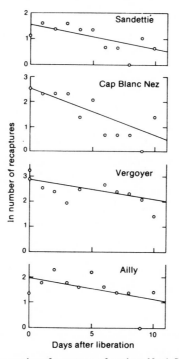

Figure 62. Regressions of log number of recaptures of southern North Sea herring on time elapsed between liberation and recapture for each of the main trawling grounds. The tagging experiment was carried out on the Downs herring stock in the winter of 1957–58. Adapted from Ancellin and Nédelèc, 1959.

Equation (97), Table 6 gives estimates of the coefficient of fishing mortality based on the regressions in Figure 62 for durations of 10 and 24 days. The numerical result would also emerge from the treatment of regressions given above. But as fish did move from one ground to another, the 10-day estimate is an underestimate of the fishing mortality, but the 24-day estimate is an overestimate, because 77 percent of the tagged fish were recovered from the position of liberation. Such biases appear in all tagging experiments, even in lakes; the solution is to use as many independent methods as possible.

Table 6.   Estimates of the coefficient of fishing mortality from the French tagging experiments with herring of the southern North Sea, 1957–58

Trawling ground	Duration of fishery	
	10 days	24 days
Sandettié	0.034	0.082
Cap Blanc Nez	0.071	0.170
Vergoyer	0.058	0.139
Ailly	0.031	0.074
Average	0.049	0.116

The herring catch in the eastern Channel in 1957–58 was 51,190 tons. If the midpoint of the estimates for the two periods is taken, then the effort involved in catching about 50,000 tons generated a fishing mortality rate of 0.08.

In another experiment, Dickie (1963) tagged cod in the Gulf of St. Lawrence, which were caught by handline and by otter trawl, a large triangular bag towed along the seabed to catch fish living near the bottom. Figure 63 shows the decline in numbers of marked cod recaptured during a period of four and one-half years after tagging. Each line represents the logarithmic decline in numbers of a length group, the smallest fish being the most abundantly distributed throughout the period.

In an extension of the argument in Equation (94), the number of fish recaptured at time t, n_1, is

$$n_1 = JHN_m \, [F/(F + X)] \{1 - \exp [-(F + X) \, t]\}, \qquad (102)$$

where H is the fraction of tagged fish that survive, and

J is the fraction of total recaptures reported.

The product JH represents Ricker's (1948) type A errors in initial losses of tagged fish due to immediate mortality and the losses represented by the failure of fishermen and others in markets and processing industries to report tagged fish, both of which could bias the estimate of F.

In a time series, as shown in Figure 63,

$$\ln n_1 = \ln JH + \ln N_m + \ln \left([F/(F + X)]\{1 - \exp[-(F + X) t]\} \right),\ (103)$$

or
$$\ln n_1 = \ln JH + \ln N_m + \ln S'',$$

where
$$S'' = [F/(F + X)]\{1 - \exp[-(F + X) t]\}.$$

With two sets of releases, N_m and N'_m, over the same time period (as, for example, two of the different length groups in Figure 63, n'_1, and n_1),

$$\ln n_1 - \ln n'_1 = \ln JH - \ln J'H'' + \ln N_m - N'_m. \qquad (104)$$

In such an equation, n_1, n'_1, N_m, and N'_m are known. If $J'H''$ is set at unity, the relative type A errors between the two sets of release groups can be determined. Such a method removes some of the biases in the estimates of F, thus pointing the way toward improved tagging techniques. In Dickie's method, the type B error, or the mortality of tagged fish due to tagging, can be estimated by differences in the slopes of the log numbers of recaptures on time. The experiment is designed to compare mortalities between batches released at the same time and place, but differing by a single factor such as length, age, or sex.

The multiple release experiment is an obvious alternative to the study of survival from a single release. In the latter the errors and biases are distributed

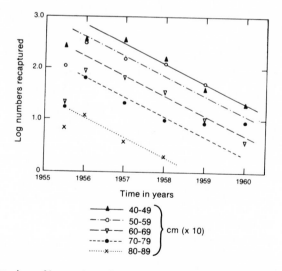

Figure 63. Regressions of log numbers of recaptures on time of different length groups of cod in the Gulf of St. Lawrence. From the analysis of data of this type, relative estimates of Ricker's type A and type B errors can be made; see text. Adapted from Dickie, 1963.

about the regression of recoveries on time, but in the multiple release they are stratified. There were several attempts to analyze multiple releases, but Jolly (1965) and Seber (1965) independently developed an experiment in which short periods of tagging and recapture are followed by long ones of no capture. It was designed for restricted seasonal fisheries in lakes, but in principle it has a more general application. Cormack (1969), in a thorough review of the statistics of mark and recapture methods, showed that it is a stochastic model. Fish recaptured more than once are classified with reference to their time of release. The form of the Jolly-Seber multiple tagging experiment is described in Table 7 (R. Jones, 1976b). The ith period represents a short one of tagging and recapture followed by a long one with neither.

At the start of the ith tagging period, there are N_i' tagged fish in the sea; during that period, N_{r_i} fish are recovered, and at the end of it N_{m_i} are released, which, of course includes N_{r_i}. N_i is the number of fish tagged before the ith period and alive at the end of that period. Hence, there were $(N_i - N_{r_i})$ fish tagged before the ith period; $N_i < N_i'$ because of losses including mortality. Table 7 shows, for example, that in the third period, 1500 fish were tagged and 156 previously tagged fish were recaptured; those 1656 fish were the total number released in period 3. Of those fish, recaptures during the fourth, fifth, and sixth periods totaled 260. Of the number of fish tagged before the ith period, $q*_i$ will be recaptured during the rest of their lives, and from N_{m_i}, r_i will be recaptured in the remainder of their lives; or, $(N_i - N_{r_i})$ is the tagged population in the ith period, $q*_i$ is the total recovery from that population, and r_i is the total recovery from the newly tagged population.

The data in Table 7 may then be used to determine $q*_i$, recaptures of those fish that were tagged before the ith period. For example, $q*_3$, recaptures of fish tagged before the third period, is the sum of recaptures during periods 4–6 of

Table 7. The structure of a multiple tagging experiment (R. Jones, 1976b)

ith period	Number of tagged fish released in ith period, N_{m_i}†	Recaptures in ith period					Total recaptures of fish released in ith period, r_i
		2	3	4	5	6	
1	1,000	60	32	21	13	3	129
2	2,060		124	82	49	12	267
3	1,656			149	89	22	260
4	1,252				188	46	234
5	2,339					193	193
Total recaptures in ith period, N_{r_i}		60	156	252	339	276	

† Includes recaptures during that period.

those tagged, N_{m_i}, in periods 1 and 2; or, looking at the table, $21 + 13 + 3 + 82 + 49 + 12 = 180$. The $q*_i$ values are as follows:

ith period	1	2	3	4	5
$q*_i$	0	69	180	188	53

$$q*_i/(N_i - N_{r_i}) = r_i/N_{m_i}, \therefore N_i = N_{r_i} + (q*_i N_{m_i}/r_i). \quad (105)$$

Then

$$S_{(i/i+1)} = r_i(N_{m_{i+1}} q*_{i+1} + r_{i+1} N_{r_i+1})/r_{i+1} N_m (q*_i + r_i). \quad (106)$$

Thus, separate estimates of survival, S, are derived for each distinct period. The exploitation rate, E, is given by

$$E_i = N_{r_i}/N_i' \cong N_{r_i}/N_i, \quad (107)$$

or the ratio of tags recaptured in the ith period to the initial number of tags in the sea. We have the survival rate [exp $(-Z)$] between periods and the exploitation ratio within periods, and if we assume that the total mortality is constant within and between periods, fishing and total mortality could be separated. However, in the structure of the experiment where fishing is restricted to a short period, the assumption is unrealistic. Strictly, the Jolly-Seber method is used for estimating stock.

However, Robson's (1963) method is useful for a multiple release experiment with continuous recapture. The numbers tagged are classified by the time when last released, and so N_{m_i} represents only those newly tagged. There are N_i tagged fish at the end of the experiment, and $q*_i$ will be subsequently recaptured; similarly, N_{t_i} fish will be tagged, of which r_i will eventually be caught again.
Then

$$q*_i/N_i = r_i/N_{m_i} \quad (108)$$

and

$$S_{(i/i+1)} = N_{i+1}/(N_i + N_{m_i})$$
$$= (N_{m_{i+1}} q*_{i+1} r_i)/N_{m_i} r_{i+1} (q*_i + r_i). \quad (109)$$

For any one period,

$$N_{r_i} = F_i N_i[1 - \exp(-Z_i)]/Z_i,$$

which is the usual catch equation, and

$$F_i = Z_i N_{r_i}/N_i [1 - \exp(-Z_i)]. \quad (110)$$

Here it is assumed that $S_{(i/i + 1)} \cong \exp(-Z_i)$, which is reasonable with continuous recapture.

The choice between single and multiple tagging experiments really depends on the chance of recapture for subsequent release. The multiple tagging systems were developed particularly in lakes, where tagging and recapture may be restricted in season and where the chance of recapture in a small population may be high. The single release was developed in the sea, where the chance of recapture was relatively low and large numbers of tags were therefore liberated at the same time.

The Estimation of Numbers by Tagging to Estimate Fishing Mortality Indirectly. Fishes have been marked or tagged for a very long time, but Petersen (1894, 1896) was the first to tag them for the purpose of population study. The estimation of numbers, by Petersen's method, is in principle very simple. If 100 fish are tagged, N_m, and 30 are recaptured, N_r, 30 percent of the stock is killed by fishing. Leslie (1952) showed that $N_r/N_m = E$, which is an unbiased, maximum likelihood estimate. From the catch equation, $C = EP_n$, $\therefore 1/P_n = E/C = N_r/N_m C$, which is also an unbiased estimate. Ricker (1945) has given a simple example of a tagging experiment and its pertinent data, using the bluegills (*Lepomis macrochirus* Rafinesque) in Muskellunge Lake, Indiana. But

$$\bar{P}_n = N_m C/N_r, \tag{111}$$

where $\bar{P}_n$ is an estimate of stock in numbers,
          N_m is the number marked,
          N_r is the number of recaptures, and
          C is the catch in numbers.

This formulation, which is Petersen's original form, demands that $\bar{P}/N_m = C/N_r$, which uses the reciprocal of the unbiased estimate from the catch equation; stock is therefore persistently overestimated, an undesirable state of affairs. In Ricker's experiment, 140 three-year-old bluegills were tagged in

Table 8. Number of recaptures from 140 tagged bluegills and the total number caught in Muskellunge Lake, Indiana, in early June (Ricker, 1945)

Fish caught	Fortnights						Total
	2nd in June	1st in July	2nd in July	1st in Aug.	2nd in Aug.	1st in Sept.	
Traps							
Recaptures	3	0	1	0	1	n.a.	5
Total catch	35	50	21	10	12	n.a.	128
Fishermen							
Recaptures	3	9	8	2	1	0	23
Total catch	120	230	165	39	36	9	599

early June, and in the ensuing weeks he recorded both the number recaptured and the total catch by scientists, using traps, and by fishermen (Table 8). Working with these data, Bailey (1951) showed an unbiased estimate of this bluegill stock to be

$$\hat{P} = N_m (C + 1)/(N_r + 1) = (140 \times 728)/29 = 3515. \tag{112}$$

A notable estimate of numbers with a multiple tagging experiment was made by Schaefer (1951) with the Petersen method on the sockeye salmon in the Fraser River, British Columbia. Table 9 shows the recoveries on the upstream migration from tags released at Harrison Mills.
Schaefer wrote:

$$P_n = \Sigma P_{ij} = \Sigma [N_{r_{ij}} (N_{m_i}/N_{r_i}) (C_j/N_{r_j})], \tag{113}$$

where i is the week of tagging and j the week of recovery.

A similar table of numbers can be constructed as the part within parentheses is calculated for each cell. This is the simplest demonstration of the power of a multiple tagging experiment. Stock in numbers is given by (C_j/E_i) corrected by the ratio of tags recovered in both weeks to that in the ith week.

The Petersen method demands that the ratio N_m/P remain constant during the period of recapture, when it might become distorted by recruitment, by migration, or merely by the inadequate mixture of the tagged population among the real one. Parker (1955) evaded the difficulty in forming a regression of n_r/C on time, the intercept of which estimates the ratio at the time of tagging before the distortions made themselves felt. The Jolly-Seber method, with the multiple recapture experiment, can be used readily to establish an estimate of population for each period in the tagging system. In the ith period, we recall that

$$N_i = N_{r_i} + q^*{}_i N_{m_i}/r_i,$$

and because

$$P_{n_i} = C_i N_i/N_{r_i},$$

then

$$P_{n_i} = (C_i q^*{}_i N_{m_i}/N_{r_i} r_i) + C_i, \tag{114}$$

which can be estimated for each period in the multiple tagging experiment. Analogous estimates can be derived with Robson's method.

With cohort analysis, estimates of stock can be made each year, if in retrospect. A tagging experiment, using Parker's method or that of Jolly and Seber, can yield the same result. In other words, a tagging experiment can be used to estimate stock in a given year, to be checked by cohort analysis several years hence. As estimates of fishing mortality also emerge from cohort analysis, such an experiment would provide independent estimates of stock and of fishing mortality.

Table 9. Recoveries of sockeye salmon tagged at Harrison Mills and recovered on the redds of the Birkenhead Rivers, a tributary of the Fraser River (Schaefer, 1951).

Week of recovery (j)	Week of tagging (i)								N_{r_j}	C_j	C_j/N_{r_j}
	1	2	3	4	5	6	7	8			
1		1	2						3	19	6.33
2	1	3	10	5					19	132	6.95
3	2	7	33	29	11				82	800	9.76
4			24	79	67	14			184	2848	15.48
5			5	52	77	25			159	3476	21.80
6			1	3	2	3			9	644	71.56
7			1	2	16	10	1		30	1247	41.57
8				7	7	6	5	1	26	930	35.77
9				3	3	2			8	376	47.00
N_{r_i}	3	11	76	180	183	60	6	1	(520)		
N_{m_i}	15	59	410	695	773	335	59	5			
N_{m_i}/N_{r_i}	5.00	5.36	5.39	3.86	4.22	5.58	9.83	5.00			

N_{r_j}, number recaptured in the jth week of recovery.

C_j, catch in number in the jth week.

N_{r_i}, number recaptured from the ith week of tagging.

N_{m_i}, number tagged in the ith week of tagging.

The Theoretical Study of Natural Mortality

When Petersen (1894) studied the stocks of plaice in the Skagerrak, he noted the occurrence in newly exploited stocks of very large old fish which were very lean and not very palatable, and which were called "praeste flyndere" or "Hanser." Garstang (1900–1903) also recorded the existence of such fish on the Dogger Bank, as did Templeman and Andrews (1956) off Newfoundland, where they were called "watery plaice." Greer Walker (1970) used Gray's formula,

$$R_s = k_3 \rho A_r V'^2 / 2g', \tag{115}$$

where R_s is the resistance of a rigid model, in kg,
         k_3 is a dimensionless coefficient,
         ρ is the density of sea water,
         A_r is the maximal cross-sectional area,
         V' is the velocity in m/sec, and
         g' is gravity in cm/sec/sec,

to show that resistance in kg increased with swimming speed much more sharply for larger than for smaller fish. In other words, the relative cruising speed (in length/sec) decreases with length, and the attack or escape speed decreases relatively to length. Greer Walker also showed that the diameter of the myofibrils of white muscle (used for acceleration) decreased in older fish, and thus the attack or escape speed decreases *absolutely* with size. If this is so, then it is possible that old and large fish are vulnerable to predation by smaller but more vigorous ones. The observations of old, lean, and not very palatable fish, in Petersen's words, in lightly exploited stocks may be interpreted in physiological terms: such old fish can no longer feed voraciously, may suffer predation, and are probably senescent, i.e., suffer a greater rate of mortality than middle-aged fishes, which effectively ends the life of the cohort.

The trend of mortality with age for a number of phyla was examined by Deevey (1947). The death rate was greatest in juvenile stages (or larval stages in fishes) and least in early and middle adult age, when reproductive activity was greatest, but in old age it increased again. In the sea and perhaps in fresh water, predation is probably the major component of mortality, although fish do die of diseases — sometimes dramatically, as for example the mass mortality of mackerel due to the fungus *Ichthyosporidium* off the east coast of the United States (Sproston, 1947). Most plants and animals in the sea live not on the surface of the seabed, where bacteria and viruses might concentrate and multiply, but in the midwater volumes, where such agents of disease must be dispersed. Sindermann (1970) gives an account of the effect of diseases on marine fish. Hence in the general sense the high mortality of larvae and juveniles must be predatory if only because higher trophic levels in the ecosystem depend upon them. As suggested above, it is even possible that senescent fish are eaten by smaller and more vigorous predators.

In a notable paper, Holling (1965) analyzed the processes of predation into components of time: time to search, time in pursuit, time to capture and handle, and time to digest (see Chapter 4). In Holling's formulation, the time to search is inversely proportional to numbers (and is also the reciprocal of the difference in speeds of predator and prey, which is itself modified by environmental factors). As the time spent searching is the predominant component of predation, the predatory mortality is density dependent.

A predator has characteristically an optimum size of prey, about one-hundredth of its own weight (Ursin, 1973), or between one-fifth and one-quarter of its length (so the speed difference is about twenty times). It follows that as the prey grows it passes from the search field of one predator to that of a larger but less abundant one. The search field of a predator is a function of its weight: (speed in lengths/sec $\times$ an area searched) is a function of l^3, and predator density is also, if it swims at the right speed to satisfy its daily ration. Any prey, because it is smaller, grows relatively faster than its predator; and as it grows into a new predatory field, the new predator must be very much less abundant than the old one. Hence predatory mortality must decline with age. Indeed, Pearcy (1962) showed that this is so (see Chapter 7).

If the chance of death decreases with time as the little fish grow through a sequence of predatory fields, that sequence is a function of growth, itself a function of age. If such predatory mortality is density dependent, then the density-dependent mortality is also a function of age. Harris (personal communication) in Cushing (1975a), developed such a formulation. Let

$$N_t = N_{(t-\delta t)} \exp - k^*_2 N_t \delta t \cong N_{(t-\delta t)}(1 - k^*_2 N_t \cdot \delta t + \ldots). \quad (116)$$

$$\delta t \to 0, \quad N_{(t-\delta t)} \to N_t,$$

$$\therefore N_t - N_{(t-\delta t)} = \delta N_t \cong - k^*_2 N_t^2 \cdot \delta t, \text{ for small } \delta t,$$

$$\int \frac{dN_t}{n_t^2} = - \int k^*_1 dt,$$

$$\therefore [-\frac{1}{N}]_{N_0}^{N_t} = [- k^*_2 t]_0^t,$$

$$\therefore (- 1/N_t) + (1/N_0) = - k^*_2 t \therefore N_t = N_0/(1 + N_0 k^*_1 t). \quad (117)$$

$$k^*_2 N = M_i ; N_t = N_0/(1 + M_0 t); M_0 = (N_0 - N_t)/(N_t \cdot t), \quad (118)$$

where k^*_2 is a constant.

Gulland, in Cushing (1977b), derived the same formula more simply. Ignore density-independent mortality:

$$M = b''N,$$
$$dN/dt = - b''N^2, \quad (119)$$
$$dN/N^2 = - b''dt,$$

$$\therefore 1/N = b''t + \text{constant} = b''t + (1/N_0),$$
$$\therefore N_t = N_0/(1 + b''N_0 t); \text{ let } b''N_0 = M_0. \qquad (120)$$

Density-independent mortality is formally ignored because when it occurs it is modulated in a density-dependent manner. The two derivations are of some interest. Gulland's implies that density-dependent mortality dies away with age as a function of numbers only. Harris's suggests that such a die-away with age might be a function of the predatory structure in the sea. This method was applied to the southern North Sea plaice; if the mortality of eggs and larvae is 80 percent per month (Harding and Talbot, 1973), $N_{30}/N_0 = 0.2$, and so M_0 and N_0 can be estimated. Figure 64 shows the trend of such mortality with

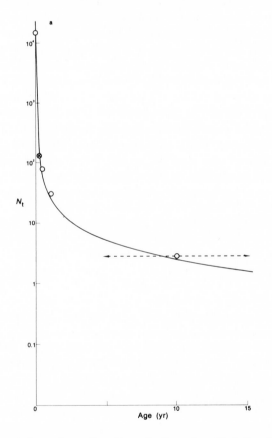

Figure 64. The trend of natural mortality with age for the southern North Sea plaice, shown by numbers, N_t, on age. The larval mortality rate is that estimated by Bannister, Harding, and Lockwood (1974); the juvenile mortality rates are those estimated on arrival and on departure from the beaches (Bannister, Harding, and Lockwood, 1974); adult mortality is that estimated from the trans-wartime year classes (Beverton and Holt, 1957). The curve fitted to the observations is that developed by Cushing (1975).

138   FISHERIES BIOLOGY

age, with the observations of total mortality of 80 percent per month in the
larval stages, 40 percent per month during the first month on the beaches, 10
percent per month during the first winter (Bannister, Harding, and Lockwood,
1974), and 10 percent per year in the adult stages (Beverton and Holt, 1957;
see below). The same method was applied to the juvenile herring off Booth-
bay Harbor and to cod material from the Baltic (Cushing, 1974). Such observations do not demonstrate the thesis, but suggest that in the future a theory of
natural mortality might embrace them more fully.

Independent Estimates of Natural Mortality. Independent estimates of natural
mortality in a heavily exploited fishery at sea have often been wartime ones,
when fishing is so light that the main cause of mortality must be natural. In
other words, it is a special case of the method of correlating mortality with
fishing intensity. Beverton and Holt (1957) have made such an estimate of the
natural mortality of the southern North Sea plaice. The year classes of 1931,
1932, 1933, and 1934 were sampled between June 1938 and April 1939 and
again between December 1945 and October 1946. The mean observed value
for the period of almost six years was $M = 0.08$, the estimate taking into
account a certain amount of fishing between April 1939 and January 1940.
Again, this value takes no account of some zero observations for some of
these year classes found after the Second World War. Beverton and Holt
suggest that a reasonable estimate from these data would probably be
$M = 0.10$. In a fuller analysis, Beverton (1964) suggested that for males
$M = 0.12$ and for females $M = 0.08$.

The North Sea herring does not live long enough for a proper estimate of
natural mortality to be made of trans-wartime year classes. However, two
estimates can be used — that from the mortalities of the Belgian spent herring
fishery, which continued during the war, and that from catch curves made in
1945 or 1946. At all times, including the war period, the Belgian spent
herring fishery exerted a low fishing intensity on the stock, as compared with
the total intensity exerted before and after the war. A catch curve yields an
estimate of mortality from a single age distribution. The logarithmic decline in
stock density from the first fully recruited age group to the oldest age group in
the single age distribution is an estimate of total mortality. There are a number
of objections to the use of catch curves, notably that differences in abundance
of age groups are largely due to year-class differences. However, they can be
used with care (see Ricker, 1958a, for their interpretations). He also gives a
number of observations on the mortality of unexploited populations in lakes,
but they are too few to establish any generalizations.

For southern North Sea herring, there is a variety of estimates of natural
mortality derived from catch curves. Such curves yielded a mortality of 0.32
in 1941–42 in the Belgian spent herring fishery (where one-half of the year
classes suffered heavier fishing before the war), 0.21 in 1942–43, and 0.13 in

1943–44. It was 0.22 in 1945 in the Boulogne fishery, and 0.17 in the Fladen fishery in 1946 (Cushing and Bridger, 1966). The average for all of these observations is $M = 0.21$. A reasonable value for the North Sea herring would be $M = 0.20$. Burd and Bracken (1965) found for the Irish Sea herring that $M = 0.15$, a figure obtained from a regression of total mortality on fishing effort. Postuma (1963), with a fuller examination of the information of the southern North Sea herring, suggested that $M = 0.08$.

There have been a number of suggestions on general grounds that natural mortality in the virgin stock increases with age. The best evidence for this proposition is given in Boiko (1964) on the zander (*Stizostedion lucioperca* [Linnaeus]) in the catchment area of the Sea of Azov. He made a number of age determinations and length measurements from remains in the Desna, Volkhov, Oka, Kama, and Neman rivers; they dated from between the third and second centuries B.C. and from the tenth to fourteenth centuries A.D. There was no real exploitation until the late eighteenth and nineteenth centuries. Boiko estimated the natural mortality in this unexploited stock as follows:

Year of life	7	8	9	10	11	12	13	14	15	16
M	0.13	0.26	0.33	0.39	0.45	0.53	0.54	0.60	0.69	1.05

Gompertz (1825) suggested that the mortality of adult animals should increase exponentially with age in order to terminate the numbers in the cohort. Beverton (1964) expressed the Gompertz law in the following form:

$$N_t = N_1 \exp -[\exp (m''_0 + m''_1 t)], \tag{121}$$

where m''_1 is the rate of increase in natural mortality with age; and

m''_0 is the initial mortality rate at eight years of age.

It can be shown that the death rate of the zander increased by 0.044/year between the ages of eight and fifteen. Cushing (1975a) attempted an analysis of the plaice age distribution assuming that $F = 0.1$; then $m''_1 = 0.09$/year.

In the Benguela current there were three sardine fisheries: at St. Helena Bay in South Africa, Walvis Bay in Namibia, and Baía dos Tigres in Angola (Davies, 1957). Between 200 thousand and 300 thousand tons of South African pilchard (*Sardinops ocellata* Pappé) and 80 thousand tons of maasbanker (horse mackerel, *Trachurus trachurus* [Linnaeus]) were caught each year. The Benguela current is very rich, and in Walvis Bay the plankton is so dense on occasion that it rots in heaps on the shoreline. To produce 1.6 thousand tons of guano annually in this area, the guano-producing penguins, gannets, and cormorants consume 43 thousand tons of pilchard and 7 thousand tons of maasbanker (Davies, 1958). The proportion of the number of fish taken by birds is not only interesting in itself, but constitutes one of the first steps in a study of the causes of natural mortality (Fig. 94). With fuller and more extensive analyses of gut contents, estimates of natural mortality might emerge.

Perhaps the most useful rule that fisheries biologists employ in an implicit way is the following: if there are many age groups in a cohort, the total mortality is not too great. Conventionally, let $N_0/N_\lambda = 100$ or 20, where N_0 is the first fully recruited age group in a cohort and N_λ the terminal one. Then ln 100 (or ln 20) $= Z$.

λ	100	20
1	4.61	3.00
5	0.92	0.60
10	0.46	0.30
15	0.31	0.20
20	0.23	0.15
25	0.18	0.12
30	0.15	0.10

If there is a given number of age groups in a cohort specified by either convention in N_0/N_λ, the greatest instantaneous total mortality can be specified, and it follows that the natural mortality is less. Independent estimates of natural mortality are not yet really available, although a number of approaches are beginning to emerge. However, the use of the rule given above in the matrices of data by age and year in cohort analysis might set limits to estimates of natural mortality.

Summary

Fishing mortality is estimated from $F = qf$, as in the Beverton and Holt regression of total mortality on fishing effort, or directly from the catch equation, as in cohort analysis (which requires an independent estimate of natural mortality). The Beverton and Holt regression has proved less successful than hoped, in general terms, although it appears to yield useful results in an unmixed fishery that has not been lightly exploited. However, if good estimates of natural mortality are not provided by the method, it often yields a reasonable measure of the exploitation ratio.

In principle, tagging experiments should provide good estimates of fishing mortality within the substock of tagged fish. However, the substock does not always represent the real one, and a number of difficulties arise in the interpretation of such experiments, even in lakes where fish are sometimes assumed to behave like billiard balls. Nevertheless, with care, reasonable estimates of the exploitation ratio can be made if the stock is fairly well exploited.

Independent estimates of natural mortality might be made under special circumstances. But our knowledge of unexploited populations is necessarily scanty, and it would take a very long time to collect any body of knowledge that would allow us to draw general conclusions. A promising avenue of study is in the physiology and ecology of old fish, which might tell us how natural mortality develops with age. It is nonetheless true that a precise separation of fishing and natural mortality remains inaccessible, and yet is one of the central problems of fisheries research.

7
Stock and Recruitment

When fishermen speak of overfishing, they believe that the run of smaller fish is due to a reduction of stock to a level at which young are no longer produced in sufficient numbers to maintain the stock. Very often, at least among demersal fish, the decline in average size is due to growth overfishing alone (i.e., the fish are caught before they reach full growth), for the number of recruits to low stock can be as much as at high stock. Recruitment per unit stock then increases with declining stock. Reasoning thus, the early fisheries biologists denied the belief of fishermen in recruitment overfishing (i.e., reduction of the magnitude of recruitment because of fishing). The fishermen's question remains, Can recruitment fail because the stock has become thinned through fishing? There are two reasons why the fishermen's question has not been properly answered. The first is that the true relation between parent stock and subsequent recruitment is always obscured by the high variation in year class strength; one is left with the impression that observations should be collected by centuries. The second reason is that the failure of recruitment might result from a single event never to be repeated in quite the same frame of circumstances.

Changes in Catches in Some Fish Stocks

Some pelagic fisheries have collapsed sharply and often without explanation. Some of the dramatic changes in catch are shown in Figure 65. Over a period of half a century, catches of the Japanese sardine (*Sardinops melanosticta* [Temminck and Schlegel]) varied by a factor of fifteen, and changes in catches of the California sardine were of two orders of magnitude (Yamanaka, 1960). Catches of the Hokkaido herring (*Clupea harengus pallasi* Cuvier and Valenciennes) (Motoda and Hirano, 1963) and of the Norwegian herring (Devold, 1963) fluctuated to the same degree. Catches of the Peruvian anchoveta increased to about 12 million tons per year between 1959 and 1970.

142

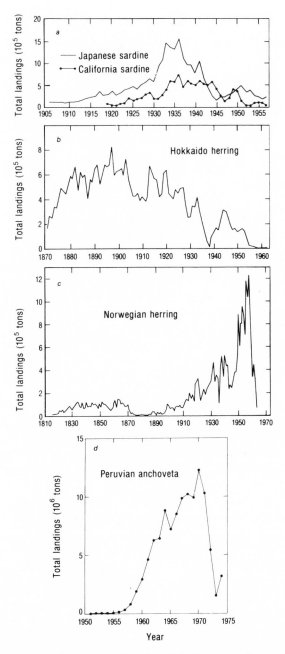

Figure 65. The catches over long periods of time for: (*a*) the Japanese sardine and the Californian sardine; adapted from Yamanaka, 1960; (*b*) the Hokkaido (or Sakhalin) herring; adapted from Motoda and Hirano, 1963; (*c*) the Norwegian or Atlanto-Scandian herring; adapted from Devold, 1963; and (*d*) the Peruvian anchoveta; adapted from Jordan, 1976.

The 1971 year class failed followed by poor recruitment in the "El Niño" years of 1972 and 1973 (see discussion of the El Niño current in Chapter 9); and from 1972–77 catches fell to about 2–3 million tons.

Changes of great magnitude have occurred over long periods. Those of the Norwegian herring stock have fluctuated for centuries, as recorded in the Icelandic sagas (Devold, 1963). Even during the 1950s, fishing mortality represented a minor proportion of total mortality, as shown from the extensive tagging results (Gulland, 1955b). During the late 1950s the stock declined, and Cushing (1968a) attributed the collapse to recruitment overfishing. The stock recovered somewhat, but in the middle 1960s Norwegian purse seiners discovered, exploited, and extinguished a shoal of Norwegian herring in the East Icelandic current, southeast of Iceland. The fishery ceased in 1967. On the other hand, there is not enough evidence to decide between fishing and natural causes for the Japanese sardine or the Hokkaido herring stocks. Figure 65 documents the range of fluctuation in the pelagic stocks, which is well known to the fishermen. Traditionally, the fishermen expect the herring-like fishes to appear and disappear. The causes of the violent variation are quite unknown save that they are basically changes in the magnitude of recruitment.

The failure of recruitment to the California sardine stock was very sudden. During the 1950s, a controversy in California revolved around the question whether the failure of recruitment in 1951 was generated by heavy fishing or whether a purely natural change had occurred (Clark and Marr, 1956). Murphy (1966) contrasted the reproductive conditions in the stock when it was exploited with those when the only cause of mortality was natural, and concluded that the stock suffered from recruitment overfishing. There was apparently a single event, the complete failure of recruitment with the 1949 year class, on which to make a judgment, and the change might have been irreversible, especially if a competitive replacement of sardine by anchovy (*Engraulis mordax* Girard) took place (Ahlstrom, 1966).

Stocks of sardines off South Africa and off Namibia also were replaced to some degree by anchovies (Newman, 1970), as was true of the Japanese sardine in the early 1940s (Zupanovitch, 1968). In recent years, the Japanese sardine has returned. The Plymouth herring (*Clupea harengus harengus* Linnaeus) stock collapsed in the 1930s with no recruitment after the year class of 1926; possibly the stock failed in competition with the stock of pilchards that succeeded it. The Plymouth herring stock was not heavily exploited, and its replacement was a natural one (Cushing, 1961). But when pelagic and opportunistic stocks are reduced to low levels by fishing, they may be replaced, as in the case of the California sardine. The pelagic stocks suffer great changes in catch from recruitment overfishing and may be replaced by pelagic competitors, particularly after heavy exploitation.

Beverton (1962) has calculated the trends in catches for certain North Sea stocks of demersal fish in a fairly long time series (Fig. 66). The North Sea cod catches remained steady for 51 years. Turbot (*Scophthalmus maximus* [Linnaeus 1758]) and plaice catches have increased slightly. Sole catches, however, have increased by ten times, and the haddock catches have fallen by an appreciable amount. During the 1960s a most remarkable change occurred. In 1962 a haddock year class was hatched which was twenty-five times larger than the preceding average, and year classes almost as large were hatched in 1967 and 1974. The recruitment to the whiting (*Merlangius merlangus* [Linnaeus]) stock increased to almost the same degree in the same years. Cod year classes were high in the years 1963, 1964, 1965, 1966, 1969, and 1970; Dickson, Pope, and Holden (1974) correlated recruitment to the North Sea cod stock with temperature and suggested that the gadoid outburst was related to cooler conditions, as the climate had become colder since the mid-1940s. Figure 84 (see below, Chapter 8) shows the increase in stock of all the North Sea gadoid species, the rate of which was common to all (Cushing,

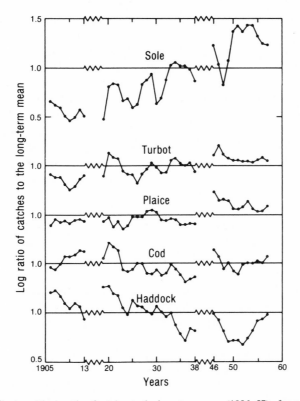

Figure 66. The logarithmic ratio of catches to the long-term mean (1906–57) of certain demersal species in the North Sea; the logarithmic ratio is used only to reduce the scale. Adapted from Beverton, 1962.

1980). Figure 66 shows that the catches of both cod and haddock were higher in the first two decades of the century, before the climate ameliorated in the 1920s and 1930s. As part of the general recovery of the gadoid stock, haddock returned to the southern North Sea and the English Channel in the 1960s, and in the Irish Sea the gadoid species recovered much as they did in the North Sea (Brander, 1977).

At a symposium in 1974 in Aarhus, Denmark, organized by the International Council for the Exploration of the Sea to discuss the gadoid outburst, other explanations were put forward. The North Sea herring stock declined between 1955 and 1968 from south to north (Burd, 1978); because it was a large stock yielding nearly one million tons in catch before collapse, the gadoids may well have taken the food released by the herring. In other words, the gadoid outburst represented a massive switch within the ecosystem under the pressure of heavy fishing, and was not an independent event associated with climatic change. No choice can yet be made between the two hypotheses, but the problem lies in the nature of the stock and recruitment relationship — how it can accommodate very large changes in numbers and whether links with other species are important. Some stocks of pelagic fish change suddenly and dramatically, but certain stocks of demersal fish rise or fall slowly for about half a century. There are no differences in larval habit or spawning behavior between the two groups of fish. The only difference is really one in growth: the pelagic fish grow quickly and demersal fish grow slowly. Pelagic fish tend to be small, and demersal ones tend to be large. The differences in growth are expressed in fecundity, large demersal fish being more fecund than small pelagic ones. The fecund stocks change slowly with time, as shown in Figure 66, but the less fecund ones suffer the dramatic changes as fisheries appear and disappear.

The Nature of the Problem

Recruitment to the East Anglian herring stock varies by factors of from 3 to 5 (Cushing and Bridger 1966); that for the arctic cod stock varies by about an order of magnitude (Garrod, 1967); and that of the North Sea haddock may be considerably greater (Sahrhage and Wagner, 1978). Figure 67 shows series of recruitments in time for a number of fish stocks in detail. Between 1887 and 1953, the year class strengths of the Karluk River sockeye salmon declined steadily. The recruitments to the herring stocks increased or declined with time, whereas those of gadoid stocks in four areas of the North Atlantic varied about a mean (except for the collapse of the Georges Bank haddock, not shown in Figure 67, in 1965). The variation is high, and most of it is a direct response to the environment. The arctic cod lays more than 4 million eggs, from which only two fish survive long enough to reproduce as adults. The implication is that there is a very fine adjustment of mortality, high stock numbers yielding about the same level of recruitment as low stock. In the

147

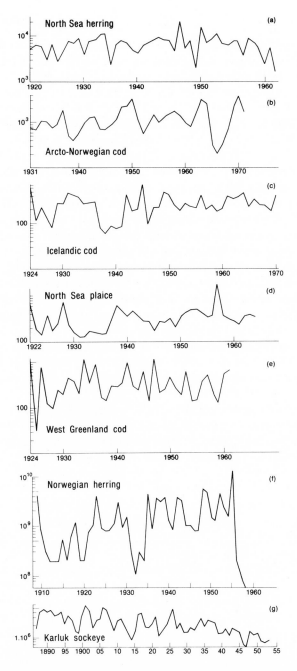

Figure 67. The variability of recruitment in numbers in long-time series for the following stocks: (a) North Sea herring; (b) Arcto-Norwegian cod; (c) Icelandic cod; (d) North Sea plaice; (e) West Greenland cod; (f) Norwegian herring; (g) Karluk River sockeye salmon. Recruitment is scaled in logarithms which compare variability between stocks. From Garrod, personal communication.

148   FISHERIES BIOLOGY

southern North Sea plaice, Beverton (1962) could establish no relationship
between stock and recruitment for a series of 26 observations. The lack of
correlation reflects the high variability of recruitment but does not establish
independence between the variables. Figure 68 shows the index of survival
(recruitment as a proportion of the stock) related to the weight of the plaice
stock (Beverton, 1962). The ratio of maximum to minimum is about 20 to 1
for the survival-rate index for a range of about 10 to 1 in the total adult
population weight. If survival is higher at low stock levels, a density-
dependent mortality operates between hatching and recruitment. Consequently the stock has a mechanism allowing it to recover from setback or
even from disaster; such a mechanism is called "compensatory." A
mechanism of quite opposite character generates a "depensatory" mortality
that increases at low stock and presents considerable danger of stock collapse.
Indeed, in some forms of depensation (Clark, 1976) the changes in stock may
be irreversible; they have not been shown in any fish stock, but the collapse of
the California sardine, like that of the Atlanto-Scandian herring, took place
very quickly. Another form of compensation is shown in the inverse relationship between the stock of pilchards in the western Channel and subsequent
recruitment (Cushing, 1961). This stock is virtually unfished and is more
abundant than the plaice stock of the southern North Sea. This fact suggests
perhaps another form of compensatory expression — where density-
dependent mortality among young fish is high enough at high stock levels to
reduce the total recruitment. The problem of stock and recruitment is a
difficult one because of the nature of this compensatory mechanism. But it is of
very great importance, because it is perhaps at the root of the great fluctuations in the fisheries. Hence, its nature should be examined in some detail.

   Hjort (1926) has suggested that fluctuations in year classes are the result of
a "critical period" in larval life — a critical period being one of high mortality, due to starvation when the yolk is exhausted. There is not much evidence
that this particular period is in fact critical. The larvae are sampled by

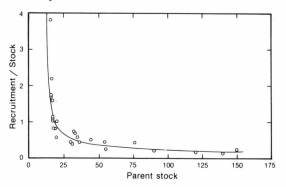

Figure 68. Index of recruit survival (or ratio of recruitment to parent stock) at different stock
levels in weight of plaice in the southern North Sea. Adapted from Beverton, 1962.

plankton nets, and the larger ones may escape. Rarely has larval mortality been separated from the two loss rates due to escape from the nets and to diffusion of larvae from the spawning center. Marr (1956) has published a larval survival curve for the Atlantic mackerel derived by Ahlstrom and Nair using Sette's information. There is no difference in Sette's original data between the day and night catches of larvae, so the larvae did not escape the net in the daytime; therefore it is likely that this curve represents both mortality and diffusion. Egg mortality is less than the early larval mortality. Later, at the time of metamorphosis, there is a step-like increase in mortality, which may be due to an increase in the escape of larvae from the nets. In a laboratory study, Farris (1960) found that the high mortality of larval sardines starts well before the yolk is exhausted, and Strasburg (1959) and Marr (1956) have recorded catches of dead larvae of sardines and other species in plankton nets which shows that larvae may die of starvation before they are eaten. Figure 69 shows that the mortality of eggs and larvae of plaice in the southern North Sea

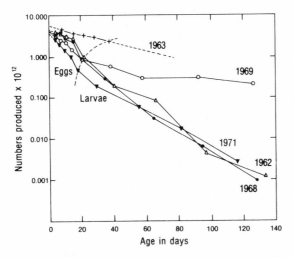

Figure 69. Mortality of eggs and larvae of plaice in the southern North Sea. Adapted from Bannister, Harding, and Lockwood, 1974.

do not differ. There is no evidence, in the studies cited, of Hjort's critical period at the time of yolk exhaustion, although there is some evidence that mortality increases significantly at the time of metamorphosis.

The Biology of Recruitment to the Parent Stock

Egg Stages. In the large populations of those species that support commercial fisheries, eggs are all about the same size, that is, about 1 mm across; exceptions are salmon, halibut, and lumpsucker (*Cyclopterus lumpus* Linnaeus), the eggs of which are 4–5 mm in diameter, up to two orders of magnitude larger by volume. They are nearly always pelagic, except those of salmon, herring, and sandeel (*Ammodytes marinus* Raitt) (of the commercial species) which stick to the gravel on the seabed. Fecundity is a function of weight; if the eggs are of the same size, bigger fish are more fecund than smaller ones. However, fecundity may not be a simple function of weight. Martyshev (1964) has shown that as carp grow older, toward middle age, their eggs and larvae are larger, but those of the oldest fish are not quite as big:

Age	4+	5+	6+	7+	8+	9+	10+	11+	12+	17+
Mean weight of eggs, in mg	1.38	1.61	1.93	2.12	2.19	2.19	2.27	2.10	2.10	1.80
Mean weight of fry, in mg	0.94	1.04	0.98	1.18	1.26	1.66	1.20	1.20	1.10	1.10

Nikolskii (1969) quotes much evidence of the same kind for different species. Older females tend to spawn earlier in the spawning season, which may last as long as three months, and so larger eggs with more yolk are hatched first. If the larvae are hatched into a food-rich production cycle (see Chapter 9), there is an advantage in releasing the larger eggs first, with their greater chance of survival. In Table 10, part (A) compares the fecundities of analogous species living in the Atlantic and in the Pacific, and part (B) those of pike and perch in the Aral and Caspian seas (Nikolskii, 1953). From parts (A) and (B), it appears that larval and juvenile mortality tends to be greater in the Pacific than

Table 10.   Fecundities of various species in thousands of eggs (Nikolskii, 1953)

Species or genus	Mean number of eggs per female	
(A)	Atlantic	Pacific
Mallotus	M. villosus Müller 6.2–13.4	M. villosus 15.3–39.9
Limanda	L. limanda (Linnaeus) 80.0–140.0	L. aspera (Pallas) 626.0–1133.0
Eleginus	E. navaga Pallas 6.2–63.0	E. gracilis (Tilesius) 25.0–210.0
Hippoglossoides	H. platessoides (Fabricius) 240.0–370.0	H. elassodon Jordan & Gilbert 211.0–241.0
Scomber	S. scombrus Linnaeus 350.0–450.0	S. japonicus Houttuyn 400.0–850.0
Gadus	G. morhua Linnaeus 570.0–930.0	G. macrocephalus Tilesius 170.0–600.0
Engraulis	E. engrasicholus (Linnaeus) 30.0	E. japonicus Schl. 35.0
(B)	Aral Sea	Caspian Sea
Pike, 30–40 cm	8.3	17.6
Perch, 16–20 cm	18.3	32.9

in the Atlantic (but not for *Hippoglossoides* or cod) and greater in the Caspian Sea than in the Aral Sea. The important point is that the difference is not restricted to one species. Again, the stocks are probably not larger in the Pacific than they are in the Atlantic. Perhaps the greater numbers of larval fish support a larger number of predators. Blaxter and Hempel (1963) have shown persistent differences in fecundity and egg size between many herring stocks in the Northeast Atlantic, which implies differences in egg or larval mortality between regions of larval drifts. Bagenal (1973) reviews information on variation in fecundity of plaice, long rough dab, witch flounder (*Glyptocephalus cynoglossus* [Linnaeus]), haddock, and pike of the same weight from year to year. It is, however, uncertain whether such variations in fecundity from year to year have any adaptive significance.

When the eggs are laid in the sea, they take up water and increase their volume considerably (up to five times in the plaice, Fulton, 1897). The gonad volume is one-fifth or one-sixth of the total volume of the fish, and the eggs have to be laid in batches in order to take up water in stages. In temperate waters, the long spawning season is the result of two processes, the batching of eggs and the sequence of spawners in time from old to young females; in subtropical waters the batching of eggs spreads the spawning season to perhaps six months, and if spawning of subtropical species is linked to food, batches may be released into rich patches of food (Cushing, 1978b).

The death rate of pelagic eggs in temperate seas is high, 5 percent per day in mackerel (Sette, 1943) and 5–6 percent per day in North Sea plaice (Harding and Talbot, 1973); that of herrings on the seabed during three weeks of development is 12 percent (Runstrøm, 1941) or 10 percent (Outram, 1958), or only about 0.05 percent per day. Harding, Nichols, and Tungate (1978) examined the mortality of plaice eggs in the Southern Bight of the North Sea for eleven years; it ranged from 1.76–12.64 percent per day; in each year, it was a log-linear function of time, which suggests that the gross differences between years are due to differences in predation. The difference of a factor of 7.5 in mortality is associated with a difference of 3.5 times in the initial numbers of eggs, but the death rate is independent of numbers. The number of predators may vary from year to year, but it is hard to imagine how they can aggregate on densities of eggs as low as 1–5/m³ in a period limited to three weeks. The lowest mortality rates (1.76 percent per day in 1963 and 3.81 percent per day in 1947) were found in the cold winters that occur about once in each decade in the Southern Bight of the North Sea, when they took twice as long to develop; hence they were vulnerable to potential predators for longer, but they died less, perhaps because the predators were also cold, with a reduced metabolic demand.

Larval Stages. During the first days of life in the sea, the larva subsists on its yolk. The larva of the Norwegian herring weighs two or three times as much as that of a Buchan herring (which spawns in the northern North Sea) and it survives for about twice as long before the yolk is exhausted, about thirty days. When the yolk has gone, larvae of either stock must have food within fourteen days; any subsequent food is of no use (Blaxter and Hempel, 1963). During its first days in the sea, the anchovy larva learns to feed. It swims intermittently and forms an S-shape laterally to discharge an attack. Its feeding success in attack in the first three days is only 10 percent, and it takes thirty days to reach 100 percent. Hence the larva can only grow successfully toward the middle of its larval life (Hunter, 1972); the Ricker-Foerster (1948) thesis that growth and mortality are linked could only work when the larvae had learned to feed and grow successfully. An anchovy larva of 1 cm in length searches about 1 liter/hour and one of 1.5 cm about 4 liters/hour (Hunter 1972), and these quantities are close to those for herring larvae (Blaxter and Hempel, 1963; Rosenthal and Hempel, 1970) and for plaice larvae (Blaxter, 1968). Lasker (1975) has shown that the first feeding anchovy gathers on thin layers of the dinoflagellate *Gymnodinium splendens* Lebour, close to the coast of Southern California. Hunter and Thomas (1974) simulated the mechanism of aggregation by anchovy larvae in the laboratory with artificial patches of the same dinoflagellate. Such mechanisms are needed because fish larvae cannot survive on an average density of food in the sea (Jones and Hall, 1973). O'Connell and Raymond (1970) showed that anchovy larvae die if they take only one nauplius/ml/day, but survive on 4 nauplii/ml/day. Wyatt (1972) analyzed the dependence of *Oikopleura* pellets/gut in plaice larvae upon the number of encounters/m. He showed that plaice depended on a low density of *Oikopleura* whereas sandeels subsisted on a high one, and so competition is limited to densities at which the distributions overlap. This short biology of fish larvae summarizes part of the extensive work now being done to illuminate the problems of stock and recruitment.

The mortality of fish larvae is sometimes difficult to determine because the baby fish can dodge the sampling nets at some point in their growth. With adequate sampling methods, it has been observed that plaice larvae die at 5 percent per day (Bannister, Harding, and Lockwood, 1974), haddock at 10 percent per day (R. Jones, 1973b), mackerel at 12 percent per day (Sette, 1943), cod at 10 percent per day (Cushing and Horwood, 1977), and herring at 4 percent per day (Graham, Chenoweth, and Davis, 1972). Larval mortality continues from that of the eggs at the same rate (Fig. 69). If that is true, there are differences in predation from year to year common to eggs and larvae. From the few observations on the plaice larvae in the Southern Bight of the North Sea, there is no evidence of stock-dependent (or density-dependent) mortality, which is a conclusion expected from a mortality common to eggs and larvae.

A good measure of larval mortality has been made by Pearcy (1962) in a study of the winter flounder in the estuary of the Mystic River in Connecticut.

Because the sizes of larvae were shown to be the same by day and by night, it is unlikely that the bigger larvae were escaping from the nets during the daytime. There is a daily loss of about 30 percent of the estuary volume, mostly near the surface. Since Pearcy found that about 85 percent of the larvae remain close to the bottom, where the eggs are laid, the daily loss by translocation to the sea is about 3 percent. The curve of larval survival with age is markedly concave, survival increasing sharply with age. From physical measurements the translocation rate (percentage removed by seaward movement) was estimated as

$$\phi = 1 - (1 - \rho'm)\psi, \tag{122}$$

where ϕ is the translocation rate,

m is the proportion of larvae in the upper 2 meters (where the main seaward translocation takes place),

ρ' is the exchange ratio per tidal cycle, and

ψ is the number of tidal cycles per day.

Since the percentage of both total loss and daily translocation loss are known, the remaining mortality can be attributed to natural mortality:

Age (in days)	Total loss	*Percentage* Loss by translocation	Loss by natural mortality
9–25	0.248	0.041	0.207
26–53	0.112	0.015	0.097

Thus, the natural mortality rate is five times the translocation rate. As numbers became reduced, the natural mortality was halved in the second period as compared with the first. It is a general principle that natural mortality which starts at such high rates must decline with age, as Cushing (1975a) has shown for the plaice on the coarser scale of years; the surprising point is that the decline in mortality rate starts in larval life, which is confirmed by recent observations on plaice larval mortality (Harding, Nichols, and Tungate, 1978).

The material from the observations on plaice larvae in the Southern Bight can be ranked in order of magnitude by years (Bannister, Harding, and Lockwood, 1974).

Eggs, stage I	Eggs, stage V	Larvae, stage 1	Larvae, stage 4
1. 1963	1. 1963	1. 1963	
2. 1962	2. 1962	4. 1969	4. 1969
3. 1968	3. 1968	3. 1968	5. 1971
4. 1969	4. 1969	2. 1962	2. 1962
5. 1971	5. 1971	5. 1971	3. 1968

The differences in numbers at hatching are small ($\times$ 1.8), but those at the last stage of larval life are considerable ($\times$ 2000); if recruitment were a density-dependent function of stock, such changes in ranking would be expected. Moreover, differences in numbers imply dome-shaped relationships between the stages sampled. If there were no compensation, the ranks should remain unchanged, as they do between stage I and stage V eggs, where differences in numbers are less than at later larval stages. The same point may be seen from the material collected on larval and adolescent cod in the Danish Belt seas (Fig. 70). Yolk-sac larvae were collected with a Petersen's young-fish trawl, and fish in age groups 0, I, and II were caught with an eel tog, which is a small trawl, not very efficient but capable of indicating gross changes in abundance. The complete data were published by Poulsen (1930a, 1930b). The material on the survival of five year classes of cod in the Baltic can be arranged in rank order of magnitude as follows:

Stock	Larval drift	0 group	II group
1924	1923	1923	1925
1927	1925	1925	1923
1923	1926	1926	1926
1925	1927	1927	1927
1926	1924	1924	1924

The same general point is made in this tabulation, and an additional point as

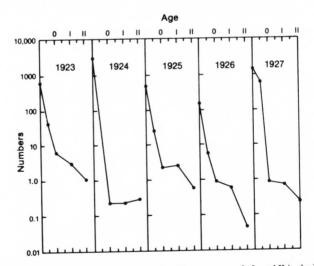

Figure 70. The larval and adolescent mortality of cod in age groups 0, I, and II in the Belt seas of central Denmark. Larvae were collected with a Petersen young-fish trawl, and fish from the 0, I, and II groups were caught with an eel tog, a form of small trawl. Data from Poulsen, 1930a, 1930b.

well — that some control can occur after metamorphosis, as will be shown below.

During the drift of eggs and larvae there is no evidence yet of a critical period such as an increase in mortality during the yolk-sac or first-feeding stages. Nor is there yet any evidence that the death rate of either eggs or larvae depends upon their numbers, that is, no evidence of density-dependent mortality generated by the aggregation of predators. However, by the Ricker-Foerster (1948) thesis, density dependence would be expressed in the time for which mortality endures according to the available food, briefly in rich food and for a long time in scarce food.

0-group Fishes. When fish larvae metamorphose, their morphology and behavior change radically. 0-group plaice turn on their sides and settle on the beaches; herring acquire fins and are caught in the waves on the shoreline; and codling swim deeper in the open sea when they start to swim like fish. Pearcy (1962) worked on the winter flounder in the Mystic River in Connecticut;

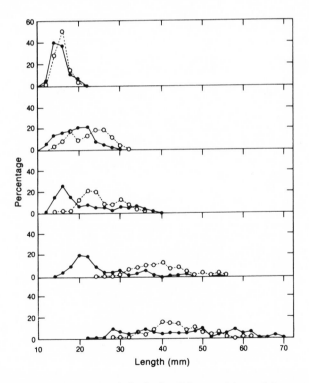

Figure 71. Changes in percentage length distribution of 0-group plaice on their nursery grounds in Firemore Bay, Loch Ewe, Scotland. Closed circles represent material from 1965; the open circles are observations from 1967. Adapted from Steele and Edwards, 1969.

Riley and Corlett (1965) examined 0-group plaice in Port Erin Bay in the Isle of Man. Macer (1967) studied the 0-group plaice population in Redwharf Bay in Anglesey, off the northwest coast of Wales. Steele and his co-workers (Steele and Edwards, 1969) analyzed the population processes in Firemore Bay in Loch Ewe in northwest Scotland. Lockwood (in press) worked on the 0-group plaice in Filey Bay in northeast England. The average mortality in these observations on flatfish in the month after metamorphosis was about 40 percent per month (1.7 percent per day), which reduced to about 10 percent per month (0.34 percent per day) six months later.

In Firemore Bay, the plaice fed on polychaete tentacles and *Tellina* siphons. Figure 71 shows how length distributions changed sharply in a very short time period. In three months, the growth of the larger individuals was considerable, but many of the smaller fish vanished, perhaps eaten by plaice (or other flatfish) of the brood one year older. The increment of growth in the short period may be an apparent one due to heavy predation. Steele and Edwards (1969) showed that the energy intake of the population remains the same, so the growth rate of individuals depends upon the mortality rate of the population. Figure 72 shows the dependence of weight increment (per unit of

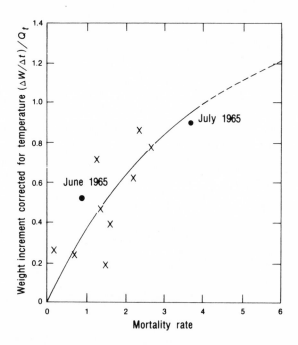

Figure 72. The dependence of weight increment of 0-group plaice, corrected for temperature — i.e., $(\Delta W/\Delta t/Q_t)$ — upon mortality in Firemore Bay, Loch Ewe, Scotland. Adapted from Steele and Edwards, 1969.

energy metabolized) on mortality rate: the greatest rate of energy transfer occurred when the death rate was highest. Not only are growth and mortality closely linked, but the proportion of growth to mortality increases with decreasing mortality rate, as time passes.

Thus growth and mortality are linked both in larvae and in 0-group fish, and there is probably a density-dependent component in both. If density dependence is a constant proportion it dies away with age, and we would expect the predominant effect to occur during larval life — but not exclusively so, as suggested in the ranking tables given above. However, the "postage stamp" plaice are attacked by cannibals a year older, among other predators.

The Theoretical Formulation of the Problem

The Malthusian principle was put by Darwin: The amount of food for each species must on average be constant, whereas the increase of all organisms tends to be geometrical, and in a vast majority of cases at an enormous ratio. The large increase is lost in the struggle for existence, and insofar as they are known, the numbers of wild populations are roughly stable. Howard and Fiske (1911) wrote that the implied density-dependent control of numbers was indicated either in fecundity or in mortality. There was a controversial literature on the nature of density dependence (Solomon, 1949; Andrewartha and Birch, 1954), but in fisheries science the principle was implicit in the use of the logistic curve, for recruitment was explicitly limited by the "carrying capacity" of the environment. Haldane (1953) and Moran (1962) expressed the necessity of density dependence as follows:

$$\Delta N = B - D' + I'' - E'', \qquad (123)$$

where B is the number of births/year,
       D' is the number of deaths/year,
       I'' is the number of immigrants/year, and
       E'' is the number of emigrants/year.
Then

$$\Delta N/N = b - d' + i - e \qquad (124)$$

in relative rates of change; if $\Delta N/N$ is to approach zero in a number of generations, some or all such rates must be functions of N.

For the reasons given in Chapter 4, density-dependent fecundity is unlikely to play a major part in control, although it may have a secondary role. If, then, mortality is the major agent in stabilization, most little fishes must be eaten by predators; no other form of mortality could satisfy rates as high as 5–10 percent per day. Predation that is the proximate cause of death may include a minority of deaths due to parasitism and disease and perhaps a majority due to varying degrees of food lack. Ricker (1954) distinguished three forms of predation: (a) the predators take a fixed number, which generates the depensatory mortality referred to above; (b) the predators take a fixed fraction of prey numbers by random encounter, a mechanism which generates the usual instantaneous coefficient of mortality; (c) the predators take all in excess of a minimum number. Some bird predators that gather rapidly from great distances may be classed in the third category, but most fish predators fall into the second, although some that attack the salmon redds or fish shoals may generate the depensatory mortality characteristic of the first category.

Because many processes in the sea must be density dependent, it is useful to distinguish them from stock-dependent ones (Harris, 1975). Between the ages of hatching and recruitment, mortality may occur at any time, and it may depend on the density at that time. A stock-dependent mechanism controls through the initial numbers and hence is effective from generation to generation, and so a density-dependent mechanism in the earlier stages of the life history is also a stock-dependent one. The major formulations may be developed as follows:

$$dN/dt = -MN,$$

where M is the instantaneous coefficient of natural mortality.

$$R = N_0 \exp [-M(t_r - t_0)], \tag{125}$$

where N_0 is the number of animals at t_0, the time of hatching, and
    R is the number of recruits at t_r, the age at which recruits accede to
        the spawning stock.
Let

$$N_0 = f^*P_e,$$

where P_e is the number of eggs spawned, and
    f^* is the number of eggs/adult. Then

$$R = f^*P_e \exp [-M (t_r - t_0)]. \tag{126}$$

This equation forms a straight line in log numbers, and we are interested in the degree of departure of reduced recruitment from it at high stock.
Let

$$dN/dt = -(M_1 + M_2N)N, \tag{127}$$

where M_2 is the instantaneous coefficient of density-dependent mortality, and M_1 is that of density-independent mortality.
Integrating,

$$R = 1/[A' + (B'/P_e)], \tag{128}$$

where $A' = (M_2/M_1) \{\exp [M_1 (t_r - t_0)] - 1\}$, and
    $B' = 1/f^* \{\exp [M_1 (t_r - t_0)]\}.$

This is the first equation of Beverton and Holt (1957), where the density-dependent mortality between t_0 and t_r becomes stock dependent at a critical stage in the life history. If mortality were constant with age, this period could occur at any age, but the death rate of larvae decreases with age (Pearcy, 1962), and to be stock dependent, the critical period must occur at an early stage in the life history. Recruitment is an asymptotic function of stock in which the initial slope is $(1/B')$, the density-independent mortality, and the asymptote is $(1/A')$, which expresses the greatest density-dependent mortality. If the density-dependent mortality is high, the asymptote is reached at a relatively low level of stock, which displays an apparent independence of recruitment from parent stock. A few observations might justify such a conclusion when the true density-dependent mortality was in fact much less.

$$M = M_0 + k_1 N_0; \qquad (129)$$

then,

$$R = N_0 \exp [- (M_0 + k_1 N_0) \, (t_r - t_0)]. \qquad (130)$$

Recall that

$$N_0 = f^* P_e,$$

$$\alpha = f^* \exp [-M_0 \, (t_r - t_0)],$$

and

$$\beta = k_1 f^* \, (t_r + t_0),$$

where α is density-independent survival, and
β is the coefficient of density-dependent mortality.
Then

$$R = \alpha P_e \exp (- \beta P_e). \qquad (131)$$

This is the Ricker equation, where the density-dependent mortality is explicitly a function of the initial numbers and hence is stock dependent. Cannibalism was the first form of stock-dependent mortality proposed by Ricker; later he expressed it as an aggregation of predators — for example, brown trout may gather on patches of salmon fry as they leave a lake. The curve of recruitment on parent stock is dome-shaped, or potentially so; the initial slope is (α) and the maximum recruitment is at $(1/\beta)$.

The equation can be derived in another way. Let

$$N_c = N_0 \exp [- \mu_1 \, (t_c - t_0)], \qquad (132)$$

where t_c is the time to reach the critical size at which a larva leaves a
predatory field, and
μ_1 is the coefficient of density-independent mortality.

$$R = N_0 \exp \{ - [(\mu_1 - \mu_2) t_c + \mu_2 t_r - \mu_1 t_0] \}, \tag{133}$$

where μ_2 is the coefficient of density-dependent mortality.
Let $(\mu_1 - \mu_2) t_c = \beta P_e$; let $\alpha = \exp(-\mu_2 t_r - \mu_1 t_0)$;

$$\therefore R = \alpha P_e \exp(-\beta P_e). \tag{134}$$

This is Beverton and Holt's (1957) second equation. It expresses the Ricker-
Foerster (1948) thesis, which states that a larva which feeds well and grows
well passes through a predatory field quickly, and then t_c is small; with more
larvae, there is less food and t_c is longer. Such a mechanism is density
dependent at any stage in the life cycle, and is stock dependent in the early
stages.

Beverton and Holt put the Ricker-Foerster thesis as follows: $t_c \, \alpha (1/W_\infty)$
α (1/food eaten) α density αN_0. Harris (1975) examined the relationship
between t_c and $(1/W_\infty)$ by differentiating dt_c/dL_∞ for small values of L,
compared with L_∞:

$$dt_c/dL_\infty = -L/KL_\infty^2. \tag{135}$$

Integrating, $t_c = L/KL_\infty$ or $t_c \, \alpha \, (1/L_\infty) \, \alpha \, (1/W_\infty^{1/3})$

$$\therefore t_c \, \alpha N_0^{1/3}, \text{ i.e., the initial distance apart of larvae in the sea;}$$

$$\therefore R = \alpha P_e \exp(-\beta P_e^{1/3}). \tag{136}$$

This brief theoretical treatment is taken from Harris (1975). The equations
used are continuous, but difference equations might be equally useful in that
the relationship can be expressed as the ratio of recruitment to parent stock for
each year of stock (Clark, 1976); this is the index of return used by W. F.
Thompson for the Fraser River stock of sockeye salmon. May (1976) demonstrated that if the dome becomes critically steep, oscillations occur leading to
"dynamic chaos," from which we conclude that the net increase rate of fishes
is low.

There are three proposed mechanisms: the aggregation of predators, cannibalism, and food limitation (either in the Ricker-Foerster sense or as simple
starvation; as will be shown below, the distinction between starvation and
food limitation under predation, in the Ricker-Foerster sense, tends to disappear). For practical purposes there is a single equation which expresses any of
the three mechanisms in a stock-dependent manner. There has been some
argument whether a dome exists in the stock and recruitment relationship, but
present evidence, as shown below, suggests that it exists in gadoids. Beverton
and Holt (1957) thought that a dome-shaped curve was unstable, but if the

right-hand limb of the dome is less than 45 percent to the abscissa, a perturbation generates a stable limit cycle. The choice of curve really depends on the nature of the mechanism, whether mortality can be shown to be a function of initial numbers or whether it endures for longer at high density.

The Ricker equation was developed for the more or less single-age stocks of the Pacific salmon fisheries. Ricker maximized Equation (131):

$$R/R_m = (P/P_m) \exp [1 - (P/P_m)], \qquad (137)$$

where R_m is maximal recruitment, and

P_m is the stock that yields R_m.

P_r is the replacement stock at which $R_r = P_r$, i.e., where the curve cuts the bisector. Then

$$R/R_r = W' \exp [a (1 - W')], \qquad (138)$$

where $W' = P/P_m$, and

$a = P/P_r$.

When $P_r > P_m$, density-dependent mechanisms predominate; when $P_r < P_m$, they are less important than the density-independent ones. Figure 73 (Ricker, 1958a) shows such curves for various conditions of recruitment per unit stock and for various rates of exploitation. The bisector is an essential part of the system because it indicates the point at which recruitment and stock are equal; as exploitation increases, the bisector shifts to the left. At a relatively low stock level there is high recruitment and at a high stock level there is low recruitment, which condition expresses the compensatory principle of density dependence.

Ricker (1950) explained the dominant cycle in the Fraser River sockeye salmon stock after the Hell's Gate landslide when he noticed that the I group fish of the dominant cycle fed on the fry of the succeeding one. Neave (1953) showed that the survival of chum salmon fry increased with increasing numbers and so the mortality was depensatory. The mechanism has been investigated by Larkin and Hourston (1964), Ward and Larkin (1964), and Larkin and MacDonald (1968), and is expressed generally in a relationship as follows (Larkin, Raleigh, and Wilimovsky, 1964):

$$(R/R_r) = W' \exp [a (W' - 1)]. \qquad (139)$$

A model was constructed by Larkin and MacDonald on the basis of five stages in the life history of the sockeye, of which the most important was the depensatory mortality. For the Adams River stock (part of the Fraser network of rivers), the model was run for many simulated years, and the subdominant cycle emerged just after the dominant one. Thus, in hypothesis, observation, and simulation, depensatory mortality exists, and the question arises whether

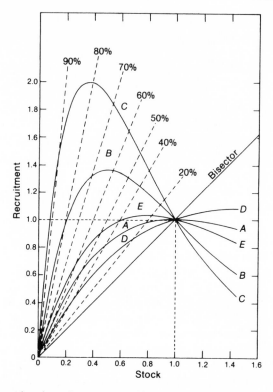

Figure 73. The stock and recruitment curves of Ricker for various conditions. The curves *A–E* represent different "stock and recruitment relationships." They pass through the bisector (where recruitment and stock are equal, and hence recruitment can replace stock) at the same point, an equilibrium point. The dashed lines represent different rates of exploitation. Adapted from Ricker, 1958a.

it occurs in marine fish stocks. Burd and Parnell (1973) have established a low concave dependence of numbers of herring larvae on their parent stock, which may indicate a form of depensatory mortality in the sea. The original explanation of the phenomenon was that herring eggs were destroyed by trawling, and destroyed most effectively at low density (Cushing and Bridger, 1966), which is of course a form of depensatory mortality.

Recently, Ricker (1973) has shown how the Ricker equation may be developed to estimate the proportion of density-dependent to density-independent mortality:

$$Z_c = \beta P_e,$$

where Z_c is the instantaneous coefficient of density-dependent mortality.

The survival rate at replacement, $S_r = (R_r/P_r)f^* = 1/f^*$,

$$\therefore Z_r = -\ln(1/f^*) = \ln f^*.$$

$$P_r = (\ln \alpha)/\beta, \therefore Z_c \text{ at replacement} = \ln \alpha; \tag{140}$$

$$\therefore Z_i = \ln f^* - \ln \alpha,$$

where Z_i is the instantaneous coefficient of density-independent mortality at all levels of stock, and

Z_r is total mortality between hatching and recruitment.

For many practical reasons Ricker's initial equation is useful because its twin derivation embraces all three of the proposed biological mechanisms. The three major dynamic processes are also expressed — the magnitude of the stock, stock-dependent mechanisms, and those due to environmental changes only. Cushing and Harris (1973) fitted this curve by least squares, i.e., by minimizing $\Sigma[R - \alpha P_e \exp(-\beta P_e)]^2$, and 95 percent confidence limits were set to the fitted curve; the method of plotting $\ln(R/P)$ on P is then avoided. This method was used because the replacement stock, P_r, is inaccessible in a multi-age stock, while the trend of natural mortality with age remains unknown. Further, in a multi-age stock the annual recruitment is only a small proportion of the stock and is equivalent to the annual loss by death. Then the stock is at its replacement point at any stable population level. Indeed, the cohorts may be considered to lead independent lives during their critical growing periods.

Stock and Recruitment Relationships and the Models That Sustain Them

Observations. Cushing (1971b) analyzed stock and recruitment relationships by groups of species (salmon, clupeids, flatfish, and gadoids) with a simple equation, $R = k^*_2 P_n^{b_1}$ (or $R/P_n = k^*_2 P_n^{(b_1-1)}$), where b_1 is an index of density dependence. The object was only to compare the index between groups, for the equation should not be used outside the range of observations. This index was inversely related to the cube root of the fecundity, which suggests that density dependence is linked to the distance apart of the eggs or larvae in the sea. More than 90 percent of the mortality between hatching and recruitment takes place during larval life, and it is reasonable to suppose that control occurs then, or not very long afterward.

The Ricker curve, or its analogue, Equation (134), represents the simplest theoretical expression which subsumes three proposed mechanisms: the aggregation of predators, cannibalism, and food limitation. Stock is expressed as number of eggs produced and recruitment as numbers at the age of recruitment to the adult stock; i.e., the process lies in the mortality from egg to recruit.

Figure 74 shows Ricker stock and recruitment curves fitted by least squares (Cushing and Harris, 1973). There is one chance in twenty that the fitted line lies outside the dashed ones. Two stocks of sockeye salmon, Karluk River (Rounsefell, 1958) and Skeena (Shephard and Withler, 1958), are represented; they are single-aged stocks, in which growth is probably density dependent. Recruitment varies about the curve by a factor of 2.5 in the Karluk stock (Fig. 74a) and by 1.5 in the Skeena (Fig. 74b), and the variability appears to decrease with declining stock. The Buchan herring is a multi-aged stock in which the first summer's growth is probably density dependent (Fig. 74c). Recruitment varies by a factor of 2.0 about the curve, and there are some high year classes at low stock, but no very poor ones. Figure 74d shows a different form of stock and recruitment curve for the chub mackerel, *Scomber japonicus* Houttuyn (Tanaka, 1974); eggs spawned by offspring are plotted on eggs spawned by parents. Both stock and recruitment are in the same units, expressing the replacement from generation to generation as in the single-aged salmon stocks. The variability about the curve is low.

Figures 74e and f show stock and recruitment curves for flatfish — Pacific halibut off Alaska and North Sea plaice (Cushing and Harris, 1973). Both are multi-aged stocks in which the variability of recruitment is rather low. Density-dependent growth has perhaps been detected in the stock of Pacific halibut at the age of three, but not among older fish (Fig. 45, Chapter 4). In Figure 74, g and h are given stock and recruitment relationships for gadoids, the Arcto-Norwegian cod (Cushing and Harris, 1973), and the North Sea haddock (Sahrhage and Wagner, 1978). Cushing and Horwood (1977) demonstrated the existence of density-dependent growth in the Arctic cod at the age of four, but not among older fish. The variation of recruitment about the curve is high (a factor of more than 25), and it is greater at low stock.

As shown above, the Ricker curve forms a dome, and the observations group themselves about it, on the left hand limb, about the middle, or across it. In the two salmon stocks the data lie on the left-hand limb of the dome; in the herring and mackerel stocks they are grouped in the same way, with perhaps a little more density dependence. The observations on the flatfish stocks lie in mid-dome, and in the gadoid stocks they lie across it. With increasing fecundity, there is greater density dependence; the gadoid dome is the most pronounced example of the trend. The estimates of average density dependence are as follows:

	$-\beta \bar{P}_e$
Pink salmon	0.61
Sockeye salmon	0.77
Herring	0.88
Flatfish	1.02
Gadoids	1.75

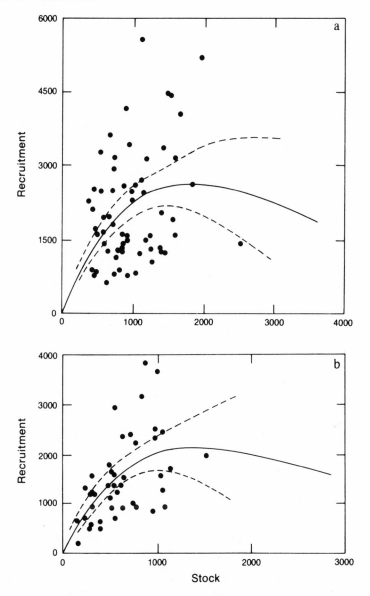

Figure 74. A gallery of stock and recruitment curves: (*a*) Karluk River sockeye salmon. Adapted from Cushing and Harris, 1973. (*b*) Skeena sockeye salmon. Adapted from Cushing and Harris, 1973. (Figure 74 continues on pages 166–68.)

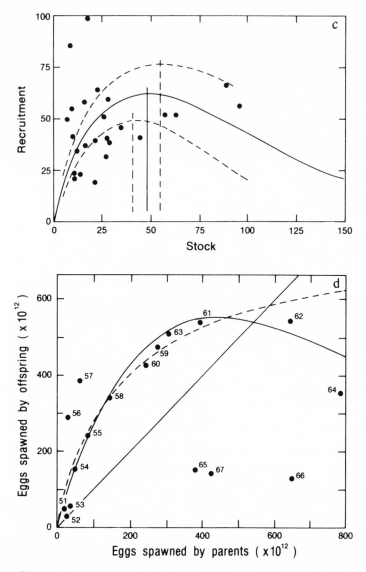

Figure 74, continued. (*c*) North Sea herring (Buchan stock); the vertical line shows the stock levels at which maximal recruitment occurs on the two error curves and on the mean. Adapted from Cushing and Harris, 1973. (*d*) Japanese mackerel; numbers beside observations are years they were made. Adapted from Tanaka, 1974.

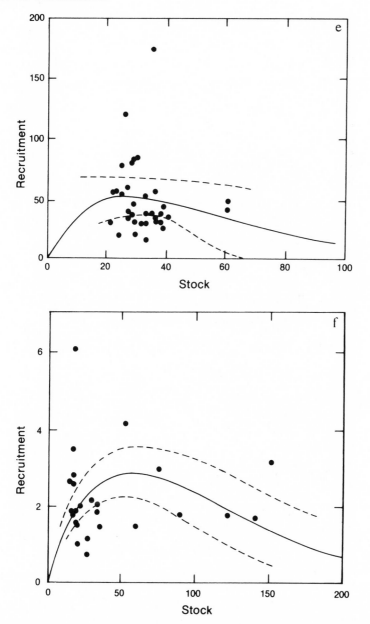

Figure 74, continued. (*e*) Pacific halibut. Adapted from Cushing and Harris, 1973. (*f*) North Sea plaice. Adapted from Cushing and Harris, 1973.

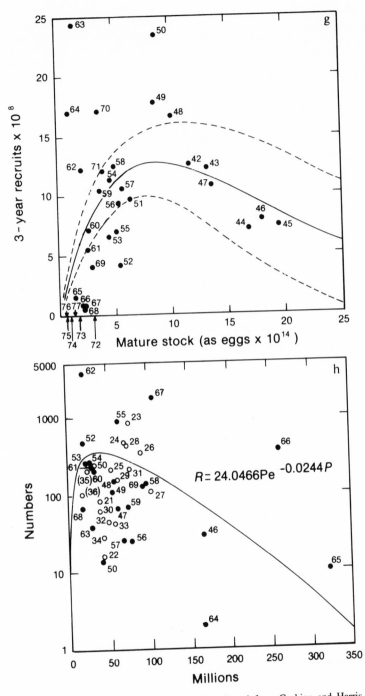

Figure 74, continued. (*g*) Arcto-Norwegian cod. Adapted from Cushing and Harris, 1973. (*h*) North Sea haddock. Adapted from Sahrhage and Wagner, 1978. Numbers beside observations are years they were made.

These estimates include a number of stocks not included in Figure 74, but listed in Cushing and Harris (1973).

In Chapter 4, density-dependent fecundity was described in the perch of Windermere. Holden (1973) thinks that density-dependent fecundity might occur in Elasmobranchii and that, because they lay few eggs, such stocks should only be exploited lightly. Conversely, the fecund cod stocks are exploited heavily because their density dependence provides them with considerable resilience. Figure 73 shows how a dome-shaped curve restores a stock more quickly than a near-linear one; a cod stock should therefore recover more quickly than a herring stock. Over the centuries, the Atlanto-Scandian herring fishery has appeared, grown, and vanished in the Norwegian Sea with a periodicity of about 110 years (Cushing and Dickson, 1976), which means that recruitment must vary by several orders of magnitude through the centuries. In contrast, the Arcto-Norwegian cod fishery in the same region has never disappeared, even if it has fluctuated considerably. In face of a similar environment, recruitment varied much less than that of the Atlanto-Scandian herring.

Harris (1975) concluded that density-dependent mortality was a function of $N_0^{1/3}$, and above it was linked to the cube root of the fecundity. The distance apart of eggs or larvae in the sea may play a crucial part in the control of marine fish populations. The parent fish may spawn at a fixed distance apart, and the density of eggs or larvae in the sea is probably a function of the number of eggs per batch spawned. If this explanation is true, a decrease in numbers in the population is associated with a reduced area of spawning, and vice versa, as shown in the decline of the California sardine population (Ahlstrom, 1966) or the expansion of the southern North Sea plaice population (Harding et al., 1978).

Models. The generation of recruitment during larval life can be simulated. In the model of R. Jones (1973b) and R. Jones and Hall (1973), fish larvae die if they fail to capture a specified number of food organisms: haddock larvae grow at 12 percent per day, die at 10 percent per day, and need at least 0.2–0.7 *Calanus* per day. The least number for survival depends on mean numbers, the search volume, and the mortality rate. As numbers vary, the minimum food requirement leads directly to a density-dependent mortality. Differences in survival of up to three orders of magnitude (i.e., in year class strength) were generated by rather small differences in the quantity of available food. An essential part of the model is that the fish larvae should grow with their food. R. Jones wrote that there would be considerable advantage in the existence of a single mechanism able to influence both growth and mortality simultaneously and hence control the balance between them in the long term.

Cushing and Harris (1973) and Cushing (1975) constructed an analogous model in which predatory mortality depended on the ratio of the cruising speed of the larvae to that of the predator. Growth rate depended on food, cruising speed on size, so predatory mortality was relatively high in poor food and relatively low in rich food; in this way both density-dependent growth and mortality were generated. The model is a development of the Ricker-Foerster thesis and links mortality and growth as did Ware (1975). Jones's model differs in that growth rate is fixed and the death of larvae is attributed to food lack, although predators may well eat the starving animals. Common to both models is the fact that density-dependent mortality is linked to food limitation during larval life.

A different form of model was made by Cushing and Horwood (1977) to investigate the form of the stock and recruitment curve. Density-dependent and density-independent mortalities were estimated from the origin to the maximum of the dome of the Ricker curve for the Arcto-Norwegian cod stock (Fig. 74g). A growth rate of 10.35 percent per day was used for the first 70 days of life, 2.4 percent per day from 70 to 168 days, and 0.71 percent per day from 168 days to three years of age. The growth rate for the rest of the life cycle was estimated with the von Bertalanffy equation, because W_{∞} can be estimated from the weight at three years of age. A competition parameter, dependent on numbers, modified growth during larval life to the limit of density dependence observed at four years of age. Mortality and growth could be linked or not. Recruitment was generated at different levels of stock; the analysis in four parts is summarized in Table 11.

Table 11.   The effects of density-dependent growth upon the nature of a stock and recruitment model (Cushing and Horwood, 1977)

Growth	Mortality	Dome	Density-dependent growth
1. Density independent	Not linked to growth	No	No
2. Density dependent	Not linked to growth	No	Yes
3. Density dependent	Linked to growth	Yes	Yes
4. Density independent	Not linked to growth; stock-dependent component added	Yes	No

Only the third combination fitted the observed presence of a dome and of density-dependent growth. But the dome was flat, and any greater mortality would have demanded a higher degree of density-dependent growth than observed. A dome could only be made by adding a component of stock-dependent mortality. The dome was generated in effect by a stock-dependent component of mortality, which may well have been cannibalism by adults — which can be stock dependent at any stage in the life history of the recruiting year class; gadoids eat young gadoids (Daan, 1975).

The observations of recruitment on parent stock may be fitted by the Ricker curve or its Beverton and Holt analogue because it expresses the three possible biological mechanisms in a convenient way. If the argument presented above on cannibalism is true, the asymptotic curve of Beverton and Holt, Equation (128), would be useful for noncannibalistic stocks. Many other expressions are conceivable, but those cited here at the least encapsulate the biological mechanisms. The models suggest that density-dependent processes in food limitation or cannibalism are sufficient, but they do not establish them nor do they deny the possibility that they could be generated by the aggregation of predators on eggs or larvae.

Summary

Many stocks of pelagic fishes have collapsed, whereas those of most demersal stocks vary to different degrees over long periods. An examination of time series of recruiting year classes shows that those of demersal stocks tend to vary about a mean, but do not always do so, and that those of pelagic stocks tend to show rising or falling trends. The survival of recruits is much less at high stock than at low stock, which is a compensatory process; on some occasions the reverse is true, and then the process is called depensatory. No decisive evidence has emerged that Hjort's critical period exists, but larval mortality may increase at metamorphosis.

The biology of fecundity — of eggs, larvae, and 0-group fishes in the sea — has been described insofar as it is known. Estimates of growth and mortality have been given for some species. The formulation of the stock and recruitment relationship in the theoretical terms of Ricker and Beverton and Holt have geen given, together with developments by Larkin and by Harris. Models to account for the process of larval growth and mortality have been described insofar as they are realistic. A gallery of stock and recruitment curves for different groups of fishes shows the high variability of recruitment about the thinly established curves; however, the density-dependent mortality is greater among the more fecund fishes.

The simplest way to account for the apparently highly variable processes by which the magnitude of recruitment is established when the cohort has reached a level of numbers "acceptable" to the adult stock is to imagine that the two processes — the generation of recruitment and the stabilization of numbers — are linked. The nature of the single process is unknown, but one might imagine that growth and mortality were both density dependent until there was enough food for the cohort and the animals could grow at their best metabolic rates.

8
Models Used in Management

The common experience of fishermen who work on demersal fish is that the harder they fish the fewer fish they catch, and the ones they do catch are smaller. In 1823, sailing smacks were landing up to 1,000 or 2,000 large turbot (*Scophthalmus maximus* [Linnaeus 1758]) after each trip in the Straits of Dover from the Sandettié Bank, or New Bank, as it was then called; by 1840, there was hardly a turbot to be found on these grounds (Alward, 1932). Fishermen know that the high catches made when grounds are first discovered cannot be repeated. They have named such grounds "Klondike" or "El Dorado." Just as the goldfields were worked out, so some grounds were apparently fished out. Such collapses have continued to the present day, exemplified by the Northeast Atlantic herring fisheries in the last two decades and the dramatic collapse of the fishery for the Peruvian anchoveta in 1971–72.

It was, however, the relatively slow decline in stock density — in the average catch of a vessel, or catch per unit of effort — with which the origin of fisheries biology was associated when such decline occurred in the last decade of the nineteenth century in the North Sea (Petersen, 1894; Garstang, 1900–1903) and in the first decade of the twentieth century off the Pacific northwest coast of North America (Thompson and Bell, 1934). The demonstration that this decline was due to fishing appeared when stock density rose in the North Sea after the first and second world wars. Changes took place, for example, in catches per unit of effort as expressed in weight of plaice in the southern North Sea during the periods both before and after the two world wars (Wimpenny, 1953; Great Britain, 1885–1965. . . . Statistical tables). The relaxation of fishing effort during each war period allowed the stock to increase in weight by many times. When peace returned, fishing was resumed and the stock fell to its prewar weight level or even lower (Fig. 75a, below).

Another characteristic of a heavily exploited fish population is the decline in mean size. This decrease in size of some demersal fish with increased fishing pressure is really a function of their growth pattern. During their lives in the fishery, such fish grow very much; if their lives are shortened by fishing, the mean sizes of the fish become much smaller, as shown in Figure 47 for herring (see above, Chapter 4). Obviously such a state can be cured by reduced fishing. The operational problem is to find the largest catch in weight for the least fishing intensity. Petersen (1894) first formulated this problem and thought that fishermen would solve it themselves because big fish fetch higher prices, but of course they can catch more small ones. It remained to Beverton and Holt (1957) to solve the problem, as will be shown below. Cushing (1972b) distinguished "growth overfishing," the condition formulated by Petersen, from "recruitment overfishing," where the magnitude of recruitment is reduced by fishing. Fishermen did not readily understand why fish became smaller as the stocks were fished harder, in growth overfishing. They believed that the breeding grounds had become disturbed, causing recruitment overfishing.

Table 12.   Estimates of potential yield and catches in 1973 by oceanic region, in thousands of tons (abridged from Gulland, 1977)

	Estimated potential		Catches in 1973
	(Gulland, 1977)	(Moiseev, 1973)	
North temperate			
Northwest Atlantic	6,400	4,800	4,485
Northeast Atlantic	13,300	11,500	11,235
Mediterranean	1,200	1,500	1,122
Northwest Pacific and Northeast Pacific	14,800	13,900	18,220
Subtotal	35,700	31,700	35,062
Tropical			
West Central Atlantic	5,500	3,400	1,405
East Central Atlantic	3,400	2,800	3,475
West Indian Ocean	8,800	3,200	1,829
East Indian Ocean	5,300	2,100	821
West Central Pacific	11,000	7,400	5,036
East Central Pacific	6,000	3,100	1,146
Subtotal	40,000	22,000	13,712
Southern temperate			
Southwest Atlantic	7,300	5,800	823
Southeast Atlantic	4,300	3,900	3,069
Southwest Pacific	2,000	1,600	300
Southeast Pacific	12,500	14,100	2,931
Subtotal	26,100	25,400	7,123
Total	101,800	79,100	55,897

It used to be thought that although certain stocks were overexploited, most were in need of development. Because the mobile distant water fleets from the industrialized nations have spread across the world ocean, and because many stocks outside the North Atlantic have not been controlled, the majority of stocks are now in need of conservation. Gulland (1977) expressed this point in Table 12, which compares two estimates of potential yield with 1973 catches by oceanic regions.

The north temperate regions are fully exploited, as probably are the southern temperate ones (catches from the southeast Pacific during the high period of the Peruvian anchoveta fishery were more than 10 million tons), but those in tropical regions are not. Although some stocks remain to be discovered, most are fully exploited and in need of conservation.

The Problem

The problems that face fisheries biologists are well illustrated in four classical case histories. The first, that of the southern North Sea plaice, is illustrated in Figure 75a. After the First World War, stock density rose by a factor of 2.0; after the Second World War, it rose by a factor of 3.3. As noted above, this was the most important early evidence indicating the effect of fishing. There are three conclusions: (1) that stock density was reduced by fishing by a factor of 2 to 3; (2) that stock density increased during the 1950s, which was partly due to stock management and partly due to a natural increase in stock; (3) stock recovered very rapidly after each world war, which implies that recruitment overfishing had not occurred and that the events were explicable as growth overfishing.

Figure 75b shows the history of the Pacific halibut fishery off British Columbia. Between 1910 and 1930, stock density (or catch per skate, a group of hooks) declined by a factor of 7, and as fishing effort increased by a factor of 3, catches fell by the same factor. After 1931, stock density increased slowly under conservation by closed season and licence; it was the first stock to recover under international agreement. There were two points of controversy: first, that the closed seasons eventually were extended so long as to be economically ridiculous (Crutchfield, 1965); second, a good year class may have appeared in 1931 (Burkenroad, 1948), and later there may have been some density-dependent effects. Today it is well recognized that a management restraint should be introduced if possible with a rising year class to minimize short-term losses to the fishermen. Skud (1975) has reassessed the measures of stock density or catch per unit effort, originally used by Thompson, and he has reexamined the Burkenroad controversy. With the long time series now available he still concludes that heavy fishing was the cause of stock decline. The two conclusions are that the fall of stock density was due to fishing as it recovered after the restraint started in 1931, and because stock

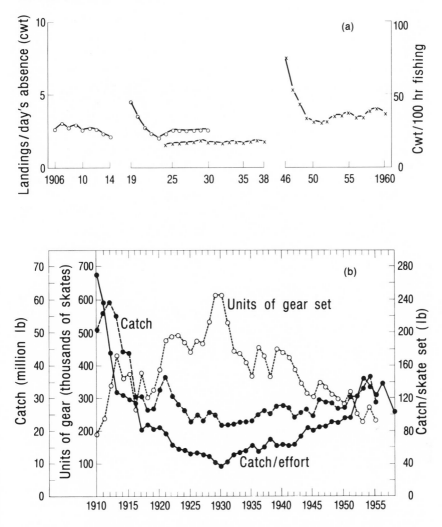

Figure 75. Long-term changes in four fisheries: (*a*) Catches per day's absence (O) and catches per 100 hours' fishing (X) in the fishery for southern North Sea plaice from 1906 to 1960. Adapted from Wimpenny, 1953; Bannister, personal communication. (*b*) Catches, fishing effort, and catches per skate (a group of six hooks on a long line) in the Pacific halibut fishery from 1910 to 1955. Adapted from Fukuda, 1962. (Figure continues on next page.)

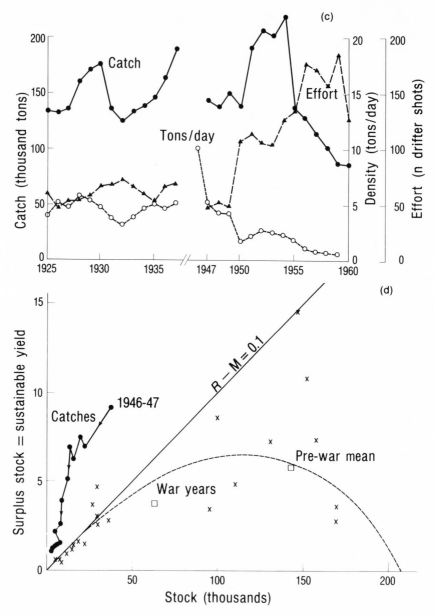

Figure 75, continued. (*c*) Catches, fishing effort, and catches per unit of effort (catch per drifter shot) in the Downs herring fishery from 1925 to 1960. Adapted from Cushing and Bridger, 1966. (*d*) Catches and surplus stock in the blue whale stock in the antarctic before and after the Second World War. The straight line estimates recruitment less mortality. Adapted from International Commission on Whaling, 1964.

density recovered quickly, the stock had suffered from growth overfishing and not recruitment overfishing.

The ancient East Anglian fishery on the Downs stock of autumn spawning herring collapsed during the decade 1955–66 (Fig. 75c). During a period of high fishing effort during the 1950s, stock density in weight fell by a factor of 30, which is very much greater than might have been expected due to loss of weight by growth overfishing. Indeed, it was shown that recruitment declined by a factor of 3 during the period (Cushing and Bridger, 1966), and we conclude that the failure of the fishery was due to recruitment overfishing.

The decline in the blue whale stock in the antarctic shows the state of recruitment overfishing most clearly (Fig. 75d). Surplus stock is plotted on stock (the Graham-Schaefer model; see below) and the observations at low stock, to the left of the figure, are fitted by the function (recruitment less natural mortality = 0.1). As stock declined, catches remained greater than surplus stock, so recruitment and stock declined sharply. Like herring, whales grow very little during their adult lives, and the collapse of the blue whale stock was due to recruitment overfishing.

The problem of overfishing is simply that too many fish are killed by fishing. Russell (1931) put it succinctly, as given in Equation (1), Chapter 1. When the catch becomes too large, R, the increment of recruitment, may be reduced, and the stock collapses due to recruitment overfishing. If, however, R is constant, catches and stock density may be reduced because the fish are caught before they have had a chance to put on the best quantity of growth, which is growth overfishing. Both conditions are cured by reducing fishing, but whereas growth overfishing merely leads to misuse of the eternally regenerating stock, recruitment overfishing will reduce it so far that the fishery is extinguished.

Surplus Production Models

Surplus production models were developed from the logistic equation (Verhulst, 1838; Pearl, 1930) used in human demographic studies:

$$dN/dt = a'N \, (1 - N/K^*), \qquad (141)$$

where N is number;

a' is the instantaneous rate of natural increase from generation to generation; and

K^* is the saturation level, or the carrying capacity of the environment.

The equation describes changes in numbers as function of the relative saturation of the environment (N/K^*) and the instantaneous rate of natural increase, a'. It is a descriptive equation which accounts for the changes in numbers with time, but provides little insight into the processes involved. Graham (1935, 1938) expressed this equation in stock in weight:

$$dP/dt = a'P \ (P_m - P)/P_m, \tag{142}$$

where P is the stock in weight;

P_m is the maximum stock in weight at the carrying capacity of the environment; and

a' is the instantaneous rate of natural increase at low stock levels.

Integrating,

$$P = P_m/\{1 + \exp [-a'(t - t_0)]\}. \tag{143}$$

In other words, the decrement in stock due to fishing is proportional to the natural rate of increase — i.e., the stock compensates naturally to the effect of fishing (or [recruitment less natural mortality] per unit stock increases relatively as stock decreases).

Then, at equilibrium,

$$FP = Y = a'P[(P_m - P)/P_m], \tag{144}$$

which expresses yield as a parabolic function of stock. This derivation is due to Ricker (1975), who showed, by differentiating yield with respect to stock, that stock at MSY (maximum sustainable yield), $P_s, = P_m/2, Y_s = a'P_m/4$, and $F_s = a'/2$, where P_s is stock at maximum sustainable yield, Y_s is yield at maximum sustainable yield, and F_s is the instantaneous coefficient of fishing mortality at which the maximum sustainable yield is maintained. Graham used this method to show that the demersal stocks in the North Sea before the First World War were overexploited.

Schaefer (1954, 1957) put the logistic equation in a different form, as shown in Equation (49), Chapter 5:

$$dP/dt = a'P(P_m - P) - FP, \tag{49, 145}$$

which depends on an independent determination of F, fishing mortality. Then

$$P = (a'P_m - F)/a' = P_m - (F/a').$$

This equation of stock (or stock density) on fishing mortality (or effort) is a negative linear expression, with slope $1/a'$ and intercept P_m.

Further,

$$Y = FP = FP_m - F^2/a', \tag{146}$$

which is a parabolic function of fishing mortality.

If $d = qP, P = d/q$ with independent estimates of F (and hence q), Equation (146) can be expressed in stock density, or catch per unit of effort. It is necessarily linear, but can be converted to a yield curve by multiplying both sides by the independently determined estimate of F, fishing mortality.

   Graham's model was only used once, but Schaefer's is the standard descriptive model which is particularly useful in stocks for which only estimates of stock density are available, or for those fishes which cannot be readily aged. The surplus production, which can be caught by fishermen, can be estimated annually without letting the stock come to equilibrium. A preliminary estimate of MSY can be obtained with observations of declining stock density through a decade or so, but such estimates can become readily biased for a variety of causes (Pope, 1976a; Horwood, 1975). Further, the increment of stock (or stock density) was initially considered as an immediate response in growth to changes in stock. As it was gradually realized from two or three decades of published age distributions that such increments were more due to recruitment than to growth, it became clear that they were delayed rather than instantaneous (as follows from the Graham-Schaefer model). However, Schaefer's original example on the yellowfin tuna (*Thunnus albacares* [Bonnaterre]) in the eastern tropical Pacific remains a good example of the method: Figure 76a shows the linear dependence of stock density (in catch per day's fishing) on fishing effort (in thousands of standard clipper days), and Figure 76b demonstrates how an MSY might be established with confidence limits. Later events suggested that the MSY was not in fact well estimated, but this figure led to the establishment of a management regime for this stock.
   Walters and Hilborn (1975) have used the Schaefer model in the following form:

$$d_1 - d_0 = a'd_0 - (a'P_m\, q)\, d_0^2 - q\, Y_t, \qquad (147)$$

where d_0 is stock density, or catch per unit of effort, in the first year; and
   d_1 is stock density, or catch per unit of effort, in the second year.
They used a stochastic treatment employing recursive methods to simulate changes in stock density in fisheries in which data had been corrected for changes in catchability. The simulated changes are shown in Figure 77 for a number of well-studied fisheries. Each simulated value is calculated from the previous one. The figure shows that the trends in stock density can be well matched once the constants a, P_m, and q have been estimated. Thus the descriptive model is used with great power. As all three constants are estimated as averages in the periods studied, understanding emerges, provided that they do not change with time.
   Such models are sometimes restricted by their parabolic nature (in the yield form). Pella and Tomlinson (1969) released the restriction by writing

$$Y = F\, [P_m - (F/a')]^{1/(m^*-1)}. \qquad (148)$$

If $m^* = 2$, we have the Schaefer model; the plot of yield on stock is symmetrical about the maximum yield as befits a parabola. If $m^* < 2$, such a yield curve is asymmetrical with the maximum displaced *toward* the origin; if $m^* > 2$, the maximum of the asymmetrical curve is displaced away from the origin

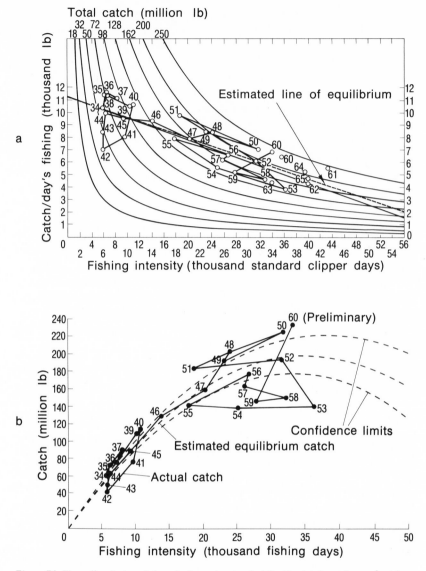

Figure 76. The yellowfin tuna fishery in the eastern tropical Pacific: (*a*) dependence of catch per day's fishing on fishing effort; (*b*) dependence of catch upon fishing effort with a Schaefer model. Adapted from Schaefer, 1957. Numbers beside observations are years they were made.

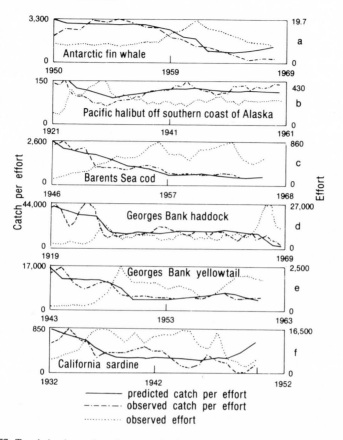

Figure 77. Trends in observed catches per unit of effort and in predicted catch per effort using recursive methods; observed trends in effort are also shown. Adapted from Walters and Hilborn, 1975.

— i.e., the stock is less affected by reduced recruitment due to fishing than when $m^* < 2$. Pella and Tomlinson have devised a program to fit Equation (148) to observations of catch and effort (because m^* is unknown); it also removes the restraint in the Schaefer model that catch per unit of effort is a linear function of effort.

There is a useful simplified procedure in stock assessment. In the Icelandic cod fishery, as sampled by British trawlers (Gulland, 1961), catch per unit of effort in one year (stock density in cwt per 100 ton-hr) tends to be inversely proportional to the average effort over the previous three years (Fig. 78). There are six age groups of fair abundance in the stock. The catch per unit of effort represents the stock density of a number of year classes; the effort averaged over three years is the mean effort modifying the stock of the six year classes. The theoretical fitting of the curve to the data demands estimates of all parameters, particularly F and M (see Chapter 5). The least-squares (solid) line fitted to the data in Figure 78 needs only estimates of stock density and of fishing effort. The equation is

$$Y/f = \delta'' - \epsilon f, \tag{149}$$

where Y/f is catch per unit of fishing intensity, and

δ'' and ϵ are constants.

The three-year averages of fishing effort, if taken in each successive year, are not independent, and neither are the corresponding estimates of stock density. The averages taken at intervals three years apart are independent, and so are the corresponding estimates of stock density. The yield equation is given simply by the parabola

$$Y = \epsilon f - \delta'' f^2. \tag{150}$$

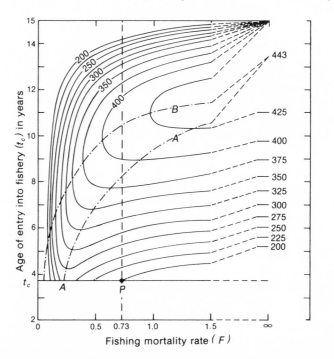

Figure 78. The catch per effort of cod at Iceland (in cwt per 100 ton-hr) and the average fishing effort over the previous three years. Any catch per effort represents the stock density of a number of year classes. The effort over three years is that average effort which has modified the stock of six year classes. The solid line has been fitted by the method of least squares; the dashed line has been fitted visually to the same data to represent the theoretical curve. Adapted from Gulland, 1961.

The dashed line in Figure 78 is a curve that has been eye-fitted to the data of catch per unit of effort on fishing effort. The distortion introduced by using a straight line rather than the curve might displace the maximum to some extent. In the Icelandic cod stocks that have been described here, the maximum yield available to fishermen was beyond the range of fishing effort. The conclusion — that the stock would stand more fishing — remains true, whether the data are fitted by eye or by a least-squares line.

This method is somewhat less restrictive than the Graham-Schaefer set of models; Garrod (1969) suggested that stock density should be expressed in logarithms, as shown in Chapter 5. Fox (1970) provides methods of estimating yield from Garrod's relationship.

The descriptive model originated in the dependence of catch per unit of effort or stock density upon effort, which was first established by W. F. Thompson (Thompson and Bell, 1934). The use of Thompson's rule led him and Michael Graham to believe that stocks should be managed by small changes in stock density. From this rule they both also believed that age determination was not needed. This expresses the real difference between the earlier descriptive models and the later analytic ones, which require independent estimates of parameters which depend upon good estimates of age distribution. Users of descriptive models can only improve their estimates with an additional annual observation which may improve their statistics marginally but will yield no further understanding. Because analytic models depend on independent parameters, they can be used without waiting for decades of observations and can be developed in their parameters.

Analytic Models

The analytic models, exemplified by the yield-per-recruit models developed by Beverton and Holt (1957), owe their origin to Russell's equation (Equation [3]) and the development of well-estimated age distributions from age determinations, combined with market sampling procedures. The development of the yield-per-recruit equation is described in Chapter 5.

The Yield Curve. The yield or catch equation given in Chapter 5 is expressed as yield per recruit, Y/R'. The reason for this is, first, that the very large fluctuation in year classes obscures the catch and fishing relationship in the data. Second, and much more important, we need to separate the relationship between catch and fishing intensity from that between stock and recruitment. The latter is a complex problem and very hard to disentangle, but it is described in Chapter 7. Here we are concerned only with the relationship between catch and fishing intensity. Model conditions are constructed by plotting Y/R' on F, the fishing mortality. As F is assumed to be proportional to fishing intensity, the curve expresses the relationship between catch and fishing.

Two such curves, for plaice and herring in the North Sea, are very different (Figure 79). For plaice (Beverton and Holt, 1957), there is a peak in Y/R' at $F = 0.2$, and at higher values of fishing mortality there is a decline in yield (Fig. 79, *top*). As fishing mortality increases, fewer fish are found in the sea, and greater numbers are caught. At the same time, the average weight of the fish declines with increased fishing mortality, each fish having less chance to put on weight. The catch is the product of the numbers caught and their weights. The one increases with fishing mortality and the other decreases. So the curve of Y/R' on F reaches a maximum value in F. If much weight is put on during the exploitable life span, the peak is sharp; if the fish do not grow at all during their lives in the fishery, there is no peak, the catch rising to an asymptote. For fish that do not grow much during their exploitable life span, the yield curve resembles that for numbers only. In Figure 79 (*top*), the dashed line at $F = 0.73$ shows the prewar yield of plaice. From the figure, it is clear that the prewar catch in the North Sea was only 80 percent of what it might have been if the fishing effort had been reduced three times. The calculations were based on an assumed mesh size of 70 mm in the cod end of the trawl. The actual mesh size was probably nearer 50 mm, and so the real state was worse than that shown in the figure.

The yield curve for herring (Fig. 79, *bottom*) is quite different (Cushing and Bridger, 1966). With increased fishing mortality, yield increases to an asymptote. There is a theoretical maximum to this curve, but in practice it is asymptotic. The biological difference between plaice and herring is that the plaice might grow twenty times in weight during its life in the fishery, whereas the herring only grows about one and one-half times. Consequently, the yield curve for the herring is very nearly that for the yield in numbers, with, however, an ill-defined maximum at very high values of F. On the rising limb of the curve for plaice, the gain in numbers is greater than the loss in weight as fishing intensity increases. On the descending limb, with harder fishing, the gain in numbers is much less than the loss in weight. In other words, the little fish are caught before they have a chance to grow.

The formal difference between the plaice and the herring concerns the nature of the growth parameters, absolutely in the magnitude of W_∞ and relatively in the value of K. The relationship between K and W_∞ tends to be inverse. If K is high, W_∞ is reached rapidly in time; if K is low, the asymptotic weight is reached slowly. In general, big fish tend to have low values of K and small fish to have high values. The difference between the growth parameters of plaice (Beverton and Holt, 1957) and herring (Cushing and Bridger, 1966) is demonstrated as follows:

	W_∞	K
Plaice	2867 g	0.095
Herring	197–224 g	0.31–0.57

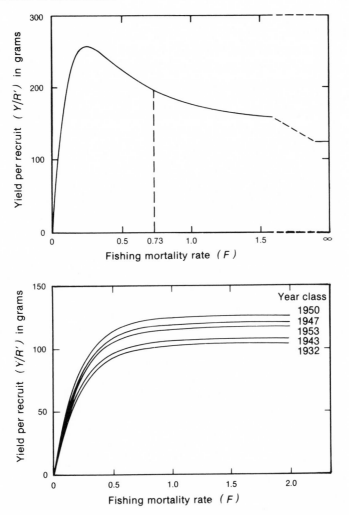

Figure 79. Top. The yield curve for plaice (as Y/R' plotted on F, fishing mortality). There is a maximum at $F = 0.2$. The dashed line at $F = 0.73$ shows the prewar condition. So an increase in total catch of 20 percent could have been obtained by reducing the number of ships by three times. Adapted from Beverton and Holt, 1957. *Bottom.* The yield curve for herring for five different year classes. No peak is found in Y/R'; in other words, with increased fishing intensity more weight of fish is landed, but only up to a certain maximum level. The difference in Y/R' in the five year classes expresses the marked effect of growth differences between them on the total catch. Relatively, such differences are much greater than those found for the plaice. Adapted from Cushing and Bridger, 1966.

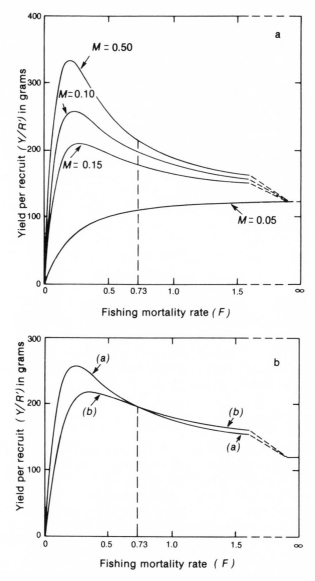

Figure 80. Yield curves for plaice calculated with different values of natural mortality and with density-dependent growth. (*a*) Yield curves calculated at a range of values of natural mortality, $M = 0.05, M = 0.10, M = 0.15, M = 0.50$. At very low values of natural mortality, the peak of the curve is sharper. (*b*) Yield curves calculated with density-dependent growth. It was assumed that L_∞ varied inversely with stock density. If growth is reduced with increased stock, the peak of the curve is reduced as might be expected. Curve (*a*) is that with constant parameters; curve (*b*) is that with density-dependent values of L_∞. Adapted from Beverton and Holt, 1957.

The Effect of Natural Mortality and Density-Dependent Growth. Different natural mortalities affect the yield curves for plaice (Fig. 80a). As noted in Chapter 6, Beverton and Holt (1957) have estimated the true natural mortality, M, for plaice (as used in Fig. 79 *top*) at 0.10. When $M = 0.05$, the peak of the curve is exaggerated as compared with that for $M = 0.10$. When M is high (0.50), the curve tends toward an asymptote, as with the herring. In such a situation, the loss in numbers by natural mortality at nearly all values of F is so high that it nullifies any excess gain in weight, which might give a peak to that curve.

The growth of fish might be density dependent (but see Chapter 4). The von Bertalanffy parameters, K and W_∞, are inversely related. Assuming K to be constant, Beverton and Holt found a decrease in L_∞ of about 20 percent for an increase in stock of six times during the Second World War for haddock in the western and central North Sea. The model used was that the asymptotic length, L_∞, varies inversely and linearly with the stock quantity. Figure 80b shows the effect of this form of density-dependent growth on the yield curves for plaice. If growth is reduced to some extent as numbers increase, the peak of the curve is reduced, which might be expected. Density-dependent growth has been detected in many freshwater fish and in the immature stages of marine fish, but it has not yet been discovered in the adult stages (but see Chapter 4). It is probably present, but differences in growth due to density cannot be readily distinguished from differences due to other causes. The models are used to show how the relation between catch and fishing might be modified by the density-dependent growth of the fish.

The Use of Yield Curves in Mesh Regulation. The practical value of these curves is in the peak. Obviously, it is wrong that fishermen should have to exploit the stock at any point beyond or to the right of the maximum. For example, in Figure 79 (*top*), to reach the point of maximum yield the prewar fishing intensity could have been reduced by three times, with the immediate effect that many companies would have become bankrupt and many fishermen driven into other employment. Therefore the change should be gradual. In the yield equation, t_c is defined as the age of entry into the fishery. It is often the age at first recruitment. If, however, t_c were to be set at a more advanced age, some small fish would be allowed to grow instead of being caught. Thus, for the same fishing effort, bigger fish are caught, and for moderate increases in the age at first recruitment the total catch is heavier. This goal could be achieved by putting larger meshes in the cod ends of the trawls, thus allowing the small fish to escape and grow.

Figure 81 is a complicated three-dimensional diagram of the yield of plaice at different mesh sizes and fishing mortality rates: t_c is shown on the ordinate as the age of entry into the fishery — in practice, this represents the mesh size of the trawl; F, fishing mortality, is set on the abscissa; values of Y/R' are described by contours. At a fixed level of F, if too big a mesh size is used in the trawl, only the biggest fish are caught, but as there are not very many really large fish, Y/R' is low. If too small a mesh size is used at that value of F, too many fish are caught before they have had time to grow. The greatest weight in the catch is found at an intermediate mesh size. This maximum yield for a given value of F rises continuously to an asymptote as a function of F and t_c, fishing effort, and mesh size.

The important point about Figure 81 is the range of choice offered in a plaice fishery, because there is an optimum mesh size for any given fishing intensity. The prewar value of $F = 0.73$ is marked on the abscissa. If t_c had been increased from age three to age nine by increasing the mesh sizes in the trawls from 70 mm to 180 mm, the possible improvement would have been a doubling of the catch for the same fishing intensity. By opening the meshes of the trawls and letting the little fish grow, the catches are increased. But it is most important to point out that this is best done gradually. A sudden and

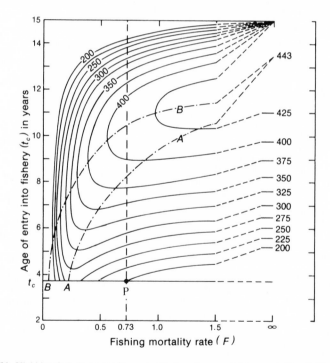

Figure 81. Yield isopleth diagram of t_c, age at first capture, on fishing mortality, F. The contours or isopleths represent the magnitudes of yield, Y/R', in g, of plaice. At $F = 0.73$, the prewar level of fishing mortality (shown at P), yield increases with increasing age at first capture to a maximum, and then it declines. The line AA' represents the condition for the largest mesh size and least fishing mortality. The line BB' represents the condition for the smallest mesh size and highest fishing mortality. Adapted from Beverton and Holt, 1957.

large increase in mesh size would cause first a decline in catch and then a recovery, because the condition could not be corrected with the existing year classes, and the recovery would have to wait for new broods to grow. What has actually happened is that the fishing mortality rate has become much reduced (0.47; Gulland, 1966) and the English fishermen at least search for the largest fish. They are effectively using a high t_c, much higher than that implied by the mesh size of 75 mm. In other words, they are carrying out their own regulations.

The examination of yield curves in management has been restricted to the trawl fisheries. Throughout the North Atlantic the trawl fisheries of all countries have been under the control of two commissions, the North East Atlantic Fisheries Commission (NEAFC) and the International Commission for the Northwest Atlantic Fisheries (ICNAF). It is a triumph of fisheries administration to have made the two commissions work and to have ensured that the catches of demersal fish from the North Atlantic are by and large well controlled, if still over exploited. In the Pacific, the International North Pacific Fisheries Commission and the Inter-American Tropical Tuna Commission carry out similar functions. Indeed, the Pacific halibut was the first stock of marine fish to be internationally regulated by the International Fisheries Commission (IFC, between Canada and the United States) since 1930.

Since the United Nations Law of the Sea Conference has been sitting, each coastal state has tended to declare an economic zone 200 miles out, which means that vessels foreign to the coastal state tend to be excluded. It is too soon to understand the effect of this trend on management, but a review of the pre-200-mile era is given in Cushing (1977a).

Heavy fishing has another effect. As fishermen increase their fishing effort, the stock in weight begins to decline accordingly. A curve illustrating this process can be constructed for herring in the southern North Sea (Fig. 82). In effect, this is a curve of catch per unit of effort, or of stock on effort ($M = 0.2$; Cushing and Bridger, 1966). As the fishermen fish harder, the stock in weight becomes less dense. An important point here is that the catch per unit of effort is not only an index of stock, but it is also an index of profit to the fishermen; Figure 82 expresses this view. With increased fishing, the number of year classes in the stock also declines. From 1956 to 1967 the southern North Sea herring provided the fishermen with essentially a one-year-class fishery (see Figure 9, which shows the age distributions from 1956 to 1960). The range of variation in the magnitude of year-class strength was high (about $\times$ 10; Cushing, 1959a), and some of this variability would be reduced if more year classes were available in the stock. With no fishing, the coefficient of variation of the annual yield due to year-class variability would be about 10 percent; when $Z = 1.0$, the coefficient of variation is about 30 percent (Cushing, 1959a). The dashed lines in Figure 82 show the coefficient of variation in catch per unit of effort, increasing with increased mortality as catch per unit of

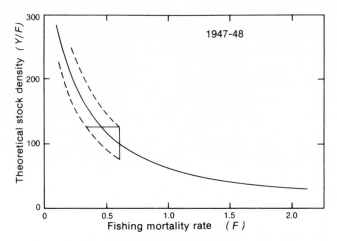

Figure 82. A curve of catch per effort (Y/F) on fishing mortality for the North Sea herring. The dashed lines represent the variability in catch per effort due to recruitment only. The vertical line represents a hypothetical low limit of catch per effort, beyond which fishermen will not go fishing. The horizontal line cuts the curve of catch per effort at a point where the low limit of catch per effort will not be reached due merely to variations in recruitment. Adapted from Cushing, 1959a.

effort falls. However, this fishery no longer exists. The last drifter left East Anglian waters in 1967; since 1977 there has been a total ban on herring fishing in the North Sea.

Clayden (1972) developed Thompson's rule (the decline of stock density with increasing fishing) with independent estimates of recruitment, fishing mortality, and natural mortality; differences in gear selectivity, availability between regions and weights-at-age were taken into account. A rough form of economic evaluation was used. A simulation model traced the changes in stock density, or catch per unit of effort, of the North Atlantic cod in ten distinct fisheries (Fig. 83). In Labrador and on the Grand Bank, the match between simulated and actual observations is poor, perhaps because of sampling problems. In other regions, particularly the Faeroes and East Greenland, the correlation between them is obviously high. This figure may be compared with Figure 76, above, which shows a simulation based on the Schaefer model. There is no difference in the quality of the match between simulated and actual observations, but Clayden's might lead to more understanding because other simulations could be attempted with changes in the independently observed parameters.

Cohort analysis has been developed to estimate the yield/recruit; then, given the recruitment, estimated independently, catch can be calculated. It will be recalled that from an array of catches at age for a number of years,

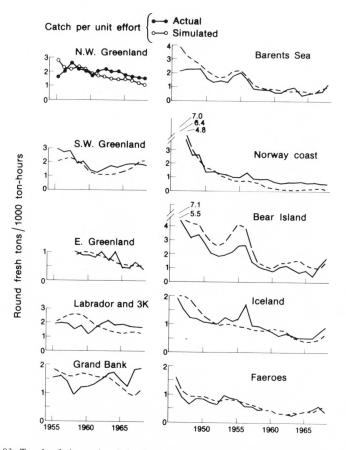

Figure 83. Trends of observed and simulated stock density of cod in a number of areas in the North Atlantic. Adapted from Clayden, 1972.

cohort analysis yields an analogous array of estimates of stock in numbers or one of estimates of fishing mortality (providing that an estimate of natural mortality is used). A sample calculation is given in Table 13. Column 2 gives the stock in the first year, including an estimate of recruiting one-year-old fish from young fish surveys; column 3 shows the percentage of maximal mortality at age, as fish recruit to the adult stock in stages; column 4 gives the mortality rates at age, with $F = 0.5$ and $M = 0.2$; column 5 gives catch in numbers at each age estimated with the catch equation; column 6 gives the mean weight at age, and column 7 the catch in weight, which is summed by age to give the total allowable catch (TAC) in tons; column 8 gives the stock in the second

Table 13. Sample calculation of total allowable catch (TAC) from cohort analysis

(1)	(2)	(3)	(4)				(5)	(6)	(7)	(8)
			Mortality							
			F = 0.5; M = 0.2							
Age	Stock in year 1	Percentage of maximal mortality	S^*F	$\exp(-S^*F)$	Z	$\exp -Z$	Catch † in numbers	Mean weight (kg)	Catch ‡ in weight (tons)	Stock § in year 2
	$N_t \cdot 10^6$	S^*					$C \cdot 10^6$	W	Y	$N_{t+1} \cdot 10^6$
1	(73)″				0.200	0.82				60
2	180	25%	0.125	0.88	0.325	0.72	19	0.18	3,420	130
3	105	50	0.250	0.78	0.450	0.64	21	0.32	6,720	67
4	48	75	0.375	0.69	0.575	0.56	14	0.48	6,720	27
5	26	100	0.500	0.61	0.700	0.50	9	0.63	5,670	13
6	10	100	0.500	0.61	0.700	0.50	4	0.76	3,040	5
7	3	100	0.500	0.61	0.700	0.50	1	0.87	870	
									26,440	

Total allowable catch (TAC) = 26,440 tons

† $C = N_t \cdot (S^*F/Z) [1 - \exp(-Z)]$.
‡ $Y = Y_n W_t$.
§ $N_{t=1} = N_t \exp(-Z)$.
″Stock of one-year-old fish estimated from surveys of young fish.

year from $N_{t+1} = N_t \exp(-Z)$. The calculation in Table 13 is essentially of the yield-per-recruit form, with independent estimates of recruitment, mortalities, and weight at age. It will be recalled that fishing mortality is estimated from the array of catches at age for many years.

There is a disadvantage in this sophisticated system; it is that the most recent estimate of fishing mortality in the cohort analysis is strictly a guess at each age, and the catches at age do not provide reliable estimates of fishing mortality until the analysis has proceeded back in the year class for three years. If the effort exerted on the stock is steady during the period, it need not matter, but good judgment is needed if effort increases during this critical period. In the Northeast Atlantic, statistics are collected slowly; catches for 1977 will be processed early in 1978 to estimate TACs in 1979, a procedure which puts an extra year into the sequence. In an ideal system of management where the stock is exploited at the maximum sustainable yield (or a little less) it might be desirable to estimate a TAC for a few years; with successive reestimates the three-year bias in F would be minimized.

There has long been controversy among fisheries biologists whether one should manage by quota or by fishing effort. The end of management is to bring fishing mortality to a required objective, loosely the maximum sustainable yield. Because $F = qf$, an effort control has appeared desirable, but it has never been reached because of the high variability of q in season and from year to year. The alternative is quota control, estimated by the cohort analysis model, which estimates the fishing mortality needed for the management objective; corresponding catch quotas are used because they can be agreed between countries.

Perhaps the simplest form of yield-per-recruit model is expressed by the decrease in mean weight with increasing exploitation:

$$W_c = E\overline{W}, \qquad (151)$$

where W_c is the weight at first capture, and

W is the mean weight of fish in the catch greater than W_c.

To get an increase in catch from a given fishery (Allen, 1953), $E > W_c/\overline{W}$. This value has been called the break-even value of E. It has been used in the ICNAF area, that is, from West Greenland to Cape Cod in the United States, in the years before the sampling systems were developed there. The value of this simplification is that, given W_c and $\overline{W}$, the range in values of E can be set at which an increase in catch can be expected. If F and M have not been determined, estimates can be made of the conditions under which catch should increase.

Any good population accountant might not describe the model as yield/recruit as put in yield curves or by cohort analysis, but for its success it demands that recruitment be constant.

Models That Incorporate the Dependence of Recruitment on Parent Stock

One of the unexpected consequences of the yield-per-recruit model is that it made recruitment overfishing a little more likely. As shown in Chapter 7, recruitment is highly variable, and any plot of recruitment on parent stock looks like a scatter diagram from which one might conclude that they were independent of each other, which is an explicit assumption of the yield-per-recruit model. Unfortunately, that scatter may conceal a true relationship which generates recruitment failure as stock is reduced under the pressure of heavy fishing. To some degree potential failure can be mitigated if recruitment is estimated as in the cohort yield-per-recruit model shown in Table 13, but it requires information on the stock-recruitment relationship. An advantage of the Graham-Schaefer model is that the differences in stock density necessarily include differences in recruitment. Provided that there are enough observations to establish the model, recruitment overfishing can be successfully avoided. However, the recovery of the Arcto-Norwegian cod stock was contrived in 1974 with the cohort yield-per-recruit model together with the stock-recruitment relationship shown in Figure 74g.

The first model that represented the stock-recruitment relationship for management purposes was that of W. F. Thompson (1945) on the stocks of Pacific salmon. The salmon are caught just offshore, in the estuaries or in the lower reaches of the rivers, and the stock that survives to spawn and die is called the escapement. Catches were regulated to maintain the escapement. The index of return is the ratio of recruiting stock to parent escapement. Such a method derives from the particular biology of the Pacific salmon (see Chapter 7).

The Ricker model of the dependence of recruitment on parent stock was developed for the stocks of Pacific salmon. Because the adults spawn only once before they die, the stock can be well estimated without the need to estimate the natural mortality from year to year as in a multi-age stock. Figure 73 (above, Chapter 7) shows a family of Ricker curves with different values of α and hence β, each normalized to the replacement value of stock, P_r, equal to the recruitment, R_r. Each curve passes through the bisector at the replacement point or equilibrium point, the condition of the virgin stock. Equation (137) in Chapter 7 expresses the Ricker model in this normalized form, based explicitly on the replacement stock. As the stock is fished, the effective bisector is shifted toward the left in Figure 73 (as shown by the dashed lines at different rates of exploitation); the catch is represented by the "surplus production" in recruitment in excess of the appropriate bisector. In this form the Ricker model provides the basis of the management of the stocks of salmon on the coast of the northeastern Pacific.

A more complex model was developed by Rothschild and Balsiger (1971) for the Bristol Bay (Alaska) salmon. It is a linear program model which includes mechanisms that describe the escapement of males and females, canning capacity, and the marginal values of the fish for the Naknek-Kvichak run. This form of model is peculiarly adapted to combining biological and economic factors because the constraints determine the mix of disparate functions in the same units.

The approach with the Ricker model is successful when applied to the stocks of Pacific salmon, but is less so for multi-age stocks. The reason is that the trend of natural mortality with age is poorly known, and R_r cannot be well estimated. If the replacement stock, P_r, is inaccessible, the bisector cannot be drawn and the method of estimating catch as "surplus production" shown in Figure 73 cannot be used. Further, it is perhaps biologically improper to use this method in a multi-age stock because the recruitment replaces only those deaths suffered by the stock during the year since the previous recruitment; in most multi-age stocks the magnitude of recruitment is only a smallish fraction of that of the parent stock.

Cushing and Harris (1973) fitted the Ricker curve by least squares for three reasons: (1) the earlier method of fitting ln $(1/P)$ on P is avoided; (2) the question of using a replacement stock in a multi-age model is evaded; (3) in its simple form (Equations [131, 134]) it expresses all the possible biological mechanisms (see Chapter 7) that have been put forward, including the Ricker-Foerster thesis that food lack generates density-dependent mortality, as shown by Beverton and Holt (1957). Figure 74 (Chapter 7) shows such curves fitted to observations on the Skeena sockeye salmon, the North Sea herring, the Pacific halibut, and the Arcto-Norwegian cod; in each, the dashed curves show the 95 percent confidence limits to the fitted curve. In other words, the fitted curve cannot lie outside the confidence limits. The same observations could perhaps be equally well fitted by the Beverton and Holt equation (Equation [128]), and the all-important low stock observations would probably be graduated in almost exactly the same way, but the possible biological mechanisms are subsumed merely in the ratio of density-dependent to density-independent mortality.

The use of the relationships shown in Figure 74 is that even if the stock at which maximum recruitment occurs cannot be determined precisely because it lies at different points on the upper and lower 95 percent confidence limits, a judgment can certainly be made by inspection. For example, the stock of Arcto-Norwegian cod should not be allowed to fall below 500 units. One of the advantages of cohort analysis is that from the array of catch-at-age data for a long period of years, estimates of stock in each year and of recruitment for each year class emerge automatically in the same units, numbers. The stock in each year raised by fecundity at age is converted to stock in eggs, and the stock-recruitment relationship emerges with recruitment in numbers. Thus the cohort analysis as used in Table 13 can be combined with the judgment made on the stock-recruitment curve to ensure best management of the stock. Such a method, with the matrix of estimated fishing mortalities, recruitments, etc., is much to be preferred to Graham-Schaefer models because changes in stock are understood in terms of the vital parameters. Cushing (1973b) described a method by which stock-recruitment curves as shown in Figure 74 can be converted to a yield curve as catch on fishing mortality. The disadvantage of this method is that such yield curves remain subject to error, although, if they are estimated at the 95 percent confidence limits to the stock-recruitment curve, the judgment of the maximum yield (in place of stock that yields maximal recruitment) can be made within established limits. There is no real difference between this method and that using cohort analysis and the stock-recruitment curve derived from it.

Natural Changes

All the models used to regulate stocks of fish make certain assumptions: (a) that natural mortality is relatively low and does not change with age; (b) that recruitment and growth may be variable, but the averages do not change from decade to decade. Very little is known of the magnitude of natural mortality, partly because of defects in the regression of total mortality on fishing intensity (see Chapter 6) but also because estimates in unexploited stocks are frequently poor. Probably the best estimate of natural mortality is that of plaice from the trans-wartime year classes (see Chapter 6); Cushing (1975a) suggested that natural mortality was a density-dependent function of age, which presupposes that most fishes, particularly young ones, die by being eaten, and implies that changes in numbers due to density-independent causes are subsequently modulated in a density-dependent manner. As fish grow they pass through a series of predatory fields, each larger than its predecessor in age; if growth depends on food in a density-dependent way, the predatory mortality may be linked to growth, itself a function of age.

As observed in Chapter 4, the ratios M/K and L_m/L_∞ (where L_m is the length at first reaching maturity) between species in the same family tend to be constant. Within the same species or within a stock, however, there might be a certain variability. But the environment could have an unexpected component. Imagine an astronomical increase in the North Sea of a predator that fed on herring. As a consequence, natural mortality of the North Sea herring would increase by many times. Obviously, such an effect would bear no immediate relationship to the growth parameters. There is also a "physiological" component of "natural" mortality, as, for example, in Bodenheimer's (1938) flies, which lived shorter lives under conditions of greater activity. In the sea, this situation could not be common; most "natural" deaths must be a form of predation because dead fish are so rarely caught. Some independent variance in predation would be an essential component to any balance between predators and prey. It is possible that the growth processes summarized by the two growth constants vary independently to some degree.

The growth rate of the North Sea herring changed considerably during the 1950s. Values of K and W_∞ in the following list are indicative of growth parameters in five year classes for the Downs stock of the North Sea herring (Cushing and Bridger, 1966):

	1932	1943	1947	1950	1953
K	0.354	0.308	0.402	0.462	0.569
W_∞	196.5	214.9	224.0	224.2	198.0

Between the observed values of K and W_∞ there is no relationship. There should be an inverse relation between K and W_∞ within the life of a fish, but it need not appear between year classes. It appears that W_∞ has increased sharply, but that K has not fallen to the extent that might have been expected.

Figure 79 (*bottom*) shows the yield curves of herring calculated for the five year classes, with K and W_∞ being used as observed in the growth-parameter list above. At a fair value of F, Y/R' can change by as much as 20 percent due to observed changes in growth parameters. It is likely that such growth changes (in addition to changes in recruitment) have had marked effects on the yields of herring in the North Sea. Growth changes have even tended to mask the changes due to fishing.

One of the most remarkable events in the history of fisheries biology was the gadoid outburst, which started in the early 1960s (Hempel, 1978) in the North Sea. The year class of haddock hatched in 1962 was 25 times larger than the average of its predecessors; very large haddock and whiting year classes also occurred in 1967 and 1974. Cod year classes were large in 1963, 1964, and 1965.

Figure 84 shows the rise in stock of all gadoid species (Andersen and Ursin, 1977), including coalfish (*Pollachius virens* [Linnaeus]) and Norway pout. The remarkable point is that all stocks increased at the same rate during the 1960s and that the range of niches is considerable, in size of fish from the small Norway pout to large cod or saithe, and in depth of water from the shallow southeastern North Sea to the deep water on the continental shelf to the north. Analogous increases occurred in the gadoid stocks in the Irish Sea (Brander, 1977), and it is possible that the changes observed in the plankton recorder network (Glover, Robinson, and Colebrook, 1972) are linked to the gadoid outburst. The quantities of many plankton organisms have been reduced considerably in the North Sea and the Northeast Atlantic (and very few have increased in the same period), and in some cases it appears that the decline was complete by the early 1960s. Because recruitment is so variable, it takes some little time to establish that a new trend is actually taking place; indeed, the gadoid outburst starting in 1962 was not really recognized by fishery biologists until the early 1970s. The scientific solution to this problem lies in the connection between recruitment and climatic factors (discussed in Chapter 9).

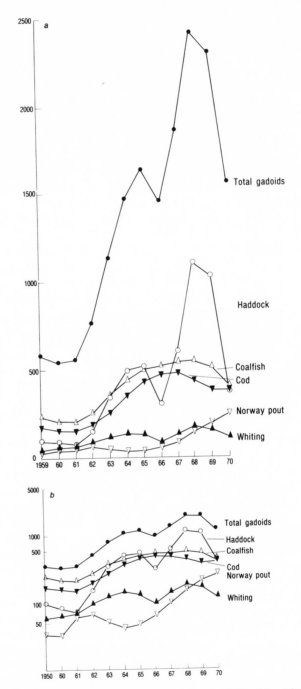

Figure 84. The gadoid outburst in the North Sea during the 1960s: (*a*) catches of haddock, coalfish, cod, Norway pout, and whiting from 1959 to 1970 (data from Andersin and Ursin, 1977); (*b*) the same information on a logarithmic scale (adapted from Cushing, 1980).

The Objectives of Management

The simple objective in the management of fish stocks is to obtain the best catches for the fishermen despite the pressures of growth and recruitment overfishing. Ricker (1946) introduced the term "maximum sustained yield." This phrase was incorporated into the Convention of ICNAF and into the 1953 treaty concerning halibut concluded by the United States and Canada. It was not defined precisely, presumably because the meaning was self-evident. Chapman, Myhre, and Southward (1962) used the term "maximum sustainable yield." Ricker (1958b) used the term in a study of yields calculated from the two stock-recruitment curves (Ricker and Beverton/Holt) under fluctuating environments and for a group of stocks. Chapman, Myhre, and Southward (1962) used the term "maximum sustainable yield" to describe the maximum of a Schaefer model of the Pacific halibut. Since then the phrase has been applied to a number of models of different forms indiscriminately. Indeed, a specific set of parameters determines one particular maximum, and another is derived from a different set. There may be many definitions in quantitative terms, but much more important is the general connotation, which any fisherman can understand, of the greatest catch that can be taken for a long period without any danger to the stock.

There is, however, an error in any determination of maximum yield, whether as confidence limits to a Schaefer estimate (see Fig. 76b) or as error in a quota derived from cohort analysis (Pope and Garrod, 1975): it is an error in fishing mortality. If the true value is less, the stock benefits, but if it is more, the added fishing mortality may reduce the stock more than intended. If a bias in method persistently generates this added fishing mortality, it could be potentially dangerous to the stock. Hence there is virtue in setting a quota at some quantity less than the maximum. Indeed, the International Whaling Commission sets its whale quotas each year at 90 percent of the MSY.

An analogous approach emerges from economic considerations. Costs are a linear function of fishing intensity and so the shape of a yield curve, decelerating toward the maximum yield with increased fishing, betrays the fact that profits (value less cost) must be greatest at some yield less than the maximum. Gulland (1968a) put this concept a little more formally as a marginal yield, where the adjective "marginal" describes an increment — i.e., the increment of yield for each increment of fishing intensity decreases as the maximum is approached with increasing fishing intensity. At the maximum of a yield curve the marginal yield is zero. A maximum economic yield was defined by Gulland (1968b) with a yield curve of value on costs; it resembles the usual yield curve in a pictorial way. It is established where the tangent to the curve is parallel to the bisector (where value = cost); the procedure is analogous to Ricker's method of obtaining a maximum yield from a stock/recruitment curve. In fishing mortality, $F_{0.1}$ is defined as that at which stock density is one-tenth of the virgin stock density, a device that approximates the fishing mortality at which the tangent to the yield curve is parallel to the bisector (Gulland, 1968b). It approximates the maximum economic yield independently of value or costs, which may differ considerably between countries. Any of these solutions may be described as an optimum sustainable yield, but, in the general sense indicated above, all are included in the meaning of maximum sustainable yield, if the word "sustainable" has an economic connotation as well as a biological one. Whatever yield curve is used, a judgment is made of how hard to fish, irrespective of whether the curve has a maximum or where it is.

So far we have assumed that stocks are managed in isolation from each other; adult stocks of halibut in the Pacific may be as isolated as possible, as they are caught on hooks. In trawl fisheries, however, catches are mixed, and it is very difficult to allocate effort by stock because the catchability coefficient depends on stock. Beverton and Holt (1957) grouped the yield curves of North Sea species in their introduction to mesh regulation; Ricker (1958a) considered the same problem theoretically with the stock-recruitment curves of four stocks. When stocks are fished in common, but are at different stages of management, particular difficulties arise: the Arcto-Norwegian haddock remains overexploited while the cod, its partner in the codend, is brought toward the maximum sustainable yield. Pope (1976a) has drawn multi-isopleth diagrams for the combined yields for two or three stocks; he has also examined the fisheries in the Gulf of Thailand, where large numbers of species are caught in the same haul in terms of the Schaefer model. This model was also used by Brown et al. (1975) to determine a maximum sustainable yield for all fin fish in the ICNAF area. Figure 85 shows a similar analysis

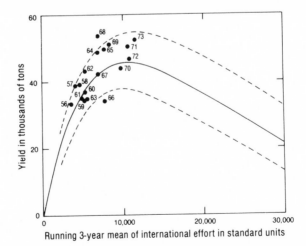

Figure 85. The use of a Schaefer model in a mixed fishery in the Irish Sea. Adapted from Brander, 1977. Numbers beside observations are years they were made.

for Irish Sea demersal stocks (Brander, 1977) which is clear because the catch per effort is in fact an index of the mixed demersal stocks and not of single species. The Schaefer model has a particular advantage in the study of mixed fisheries, provided that data are available for long enough; the advantage is that the catchability coefficient is averaged for all stocks and assumed constant with time. The single-stock approach, shown in Table 13, should yield the same result once the catchability coefficient can be partitioned between stocks.

A very much harder problem in a mixed fishery is that where biological interactions might occur — for example, antarctic whales and krill (*Euphausia superba* Dana) (Horwood, 1976). Cod eat haddock in the North Sea in large numbers (Daan, 1973), but it is not yet known what the nature of the quantitative interaction might be. If a stock is fished to a very low level, it can be replaced by another, as the northern anchovy replaced the California sardine (Ahlstrom, 1966). Andersen and Ursin (1977) have suggested that the gadoid outburst referred to above might be due to the loss of herring and mackerel stocks in the North Sea, but the existence of an outburst in the Irish Sea without loss of herring or mackerel appears to deny the hypothesis. Such problems require not only an understanding of how recruitment is generated from the parent stock, but a mastery of that problem. Chapter 7 shows how far we are from that position.

The problems of management are distinct from those of estimation; a yield curve from cohort analysis and a stock-recruitment curve based on the same material provide good data for estimation. But the judgment of what to do with this information is of a different quality and may vary from case to case. In 1974, the Arcto-Norwegian cod stock was in considerable danger due to recruitment overfishing; the recruiting year classes were probably good, and whereas a population accountant would have prescribed Draconian measures, the prospective recruitment allowed the judgment that the maximum sustainable yield could be reached by riding on the good year classes with no loss in catch or employment to the fishermen. This form of judgment by the scientists is much more important than the strict adherence to the limited quantities that emerge. This judgment is of the same nature as the more general connotation of the maximum sustainable yield in the simple words that any fisherman or politician can understand.

Conclusion

It is a truism that the success of management depends on the quality of the science that supports it. The initial steps made in science by Graham and Thompson were enough to indicate the need for management in its simplest form. This form was the restraint of effort until stock density was restored by small steps to the magnitude desired. The more developed science, based on independently determined parameters in the analytic models used first by

Ricker and later by Beverton and Holt, led to management by fishing mortality. At first that management was roughly conceived in terms of mesh regulation (as in the yield isopleth diagram) and as such was put into practice all over the North Atlantic.

But the new management based on better science led to two distinct failures. The yield-per-recruit model combined with the high variability of any stock-recruitment curve obscured the decline in herring recruitment during the late 1950s and the 1960s in the Northeast Atlantic. Scientifically, this failure was mitigated by progress in understanding the nature of the stock-recruitment relation, but the failure in management was absolute. The second failure in management was that the overall application of mesh regulation in the North Atlantic did not prevent the majority of stocks from being overexploited, although in theory it might have done. Overexploitation occurred because fishing effort, in number of ships, increased too quickly. However, by the time that overexploitation had been admitted, the method of cohort analysis had become available, by which time much more information on fishing mortality could be marshaled and transposed into catch quotas. Despite their inconvenience, such devices stabilize catches and prevent further expansion.

Today the problems of mixed fisheries, both as combined catches of disparate stocks and as samples of stocks that potentially interact in biological terms, dominate the scientific field. The quotas based on estimates of single species will generate disorder and remain suspect until both of these problems of mixed fisheries are solved. Present management is not as successful as it would be if quotas were calculated on a mixed-fishery basis and if the potential biological interactions between stocks were known.

9
Fish in Space and Time

The Ocean Boundaries

The ocean appears to be featureless. With their range of vision limited by the physical properties of seawater to a sphere with a radius of 10–100 m, fish see only a few food targets at any one time. Although some fish such as the common sole may smell their food or even hear it, most of the abundant fishes in the commercial fisheries are visual feeders. There are well-defined physical structures in the waters of the ocean, and fish populations sometimes gather there. There are three associated facts — the short range of perception, the long range of migration among some species, and the presence of shoals at the oceanic boundaries. It is possible that migrant fish use the oceanic boundaries to make their migrations. Because fish appear regularly each season at such boundaries, the fisheries do so also; this is our reason for studying them in some detail.

There are three main oceanic boundaries:

(1) isotherms that are sharp enough to generate changes in behavior and perhaps to limit the range of the stock;

(2) current boundaries against which the fish pack, perhaps for food, perhaps mechanically, or perhaps because they see them;

(3) an area of upwelling or divergence, where water from below rises to the surface, where plankton production is often intense, and where fish gather.

Fish concentrate at each of these boundaries. The causes of concentration are examined below, insofar as they affect the concepts of fisheries biology. The most important of these concepts is the idea of a unit stock, as oceanic boundaries might really be stock boundaries. In the upwelling areas are found the best conditions for the survival of larvae and juveniles which would illuminate the problem of stock and recruitment. Four problems that relate to fisheries biology are discussed: the restriction of the arctic cod by the 2° C isotherm on the seabed; the relation of herring to horizontal boundaries; the relation of tuna to current boundaries; and the production of fish in the upwelling areas.

Fish are caught where they are concentrated. They gather at some point on their migration circuit to spawn, to feed, or to take advantage of the current structure in making their migrations. The arctic cod and the Pacific tuna gather at oceanic boundaries perhaps to feed to the best advantage. Such advantage is certainly present when the North Sea herring pack against the Baltic outflow in early spring. Fisheries are established at such boundaries because the fish live there during part of their migration cycles. In the great upwelling regions, the whole life of sardine-like fishes may be geared to the food production system. The structure of the ocean is moderately well known, on the broad scale of the major ocean currents, but the lesser scales that fish might exploit on their ocean passages are not yet very well known. From the nature of migration circuits and the dependence of fish on production cycles, fisheries biologists need to know the courses of the oceanic currents.

The Arctic Cod and the 2° C Isotherm on the Bottom. It was shown in Chapter 3 (Cushing, in Richardson et al., 1959) that, in midsummer on the Svalbard shelf, the arctic cod do not appear to cross the 2° C isotherm on the bottom or

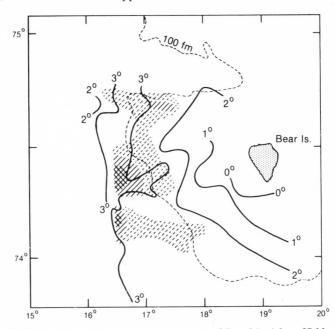

Figure 86. Cod catches and bottom temperatures west of Bear Island from 27 May to 5 June 1950. Distribution of cod catches is presented with respect to the 2° C isotherm on the bottom. Seabed temperatures were recorded at the end of each haul, and contours are of isotherms on the bottom. The differently hatched areas indicate where the highest catches of cod were obtained. Adapted from Lee, 1952.

the same isotherm in midwater during spring and autumn in the southeastern Barents Sea (Hylen, Midttun, and Saetersdal, 1961). Figures 86 and 87 contrast a survey of trawl catches and bottom temperatures made in early June 1949 with a similar one made in late May and early June 1950 (Lee, 1952). In 1950 cold water lay close to Bear Island, and in 1949 it was spread to the very edge of the Svalbard shelf. In both cruises, the best catches were found in water warmer than 2° C, regardless of depth. The same results were obtained on a number of occasions (Lee, 1952). It is worth pointing out that the catches were heavy, that British trawlers steamed 1,500 miles to crowd on the Svalbard shelf in summer to form what the fishermen called "arctic cities."

Another region where cold and warm waters conflict to the profit of fishermen is the Grand Bank. The Gulf Stream floods the southern edge of the Bank before swinging eastward into its meandering drift across the North Atlantic; the cold Labrador current covers the northern edge of the Bank and sometimes covers the major part of it (Dietrich, 1965). Templeman and May (1965) and Templeman and Hodder (1965) have described the distributions of cod and haddock by catch per unit of effort at temperature and at depth. For example, in April and May, cod are concentrated in temperatures just above 2° C in depths of about 250 fm. Sections of temperature in depth and in catch per unit

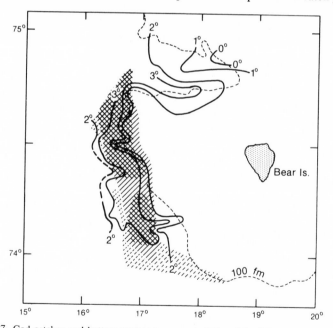

Figure 87. Cod catches and bottom temperatures west of Bear Island from 31 May to 6 June 1949. Distribution of cod catches is presented with respect to the 2° C isotherm on the bottom. Seabed temperatures were recorded at the end of each haul, and contours are of isotherms on the bottom. The differently hatched areas indicate where the highest catches of cod were obtained. Adapted from Lee, 1952.

of effort of haddock showed quite clearly that the fish avoided the intrusions of much colder water. Such effects might be expected wherever a boundary might affect the fish directly and in a physiological sense.

Cod can survive in very cold water ($-1.5°$ C), but then the fish are effectively supercooled liquids, and if they are touched in this incredible state they turn into cod-shaped blocks of ice (Sundnes, 1965). Woodhead and Woodhead (1959) showed that cod caught at sea in water cooler than $2°$ C have more chloride in their blood than fish from warmer water. Under more rigorous experimental conditions in the laboratory, Harden Jones and Scholes (1974) concluded that there was no physiological "barrier" at $2°$ C, but that the exchanges of water and chloride were reduced, perhaps because of the existence of antifreeze in the blood. It is likely that there are some physiological consequences of migrating into cold water, but the observations made at sea may have a hydrodynamic origin in that the fish live in the warmer water.

Figure 28 (above, Chapter 3) shows three echo surveys (Richardson et al., 1959) that were made within a fortnight of each other; they described the movement of cod onto the shelf, and in each survey the echo traces of cod were bounded by the $2°$ C isotherm on the bottom. On the western edge of the Svalbard shelf, there are two submarine inlets, the Størfjordrenna, south of Spitsbergen, and the Northwest Gully, north of Bear Island. The Størfjordrenna was flooded with cold water that had drifted on the bottom down the east side of Spitsbergen. The Northwest Gully was flooded with warm water that had pushed up from the West Spitsbergen current. On the first echo survey, the cold concentrations were bounded neatly by the $2°$ C isotherm. Within a fortnight, on the third survey, the cod patches had moved up the shelf from about 150 fm to about 80 fm. At the same time, the $2°$ C isotherm had moved up the shelf to the same extent. Some six weeks later a further survey (Beverton and Lee, 1964) was made over the whole of the Svalbard shelf, and the $2°$ C isotherm had spread to its eastern edge. The echo patches of cod had diffused over the area, and there were no longer any dense concentrations worth fishing.

Since dispersion would ordinarily be expected after a movement of three or four hundred miles, the concentration actually found might at first seem surprising. The question is whether the cod were thrown mechanically onto the shelf by the upward movement of the warmer water or whether they were concentrated against the cold-water barrier by some physiological factor. The estimated quantity shown in Figure 28 might be a fair fraction of that spawning stock which had migrated back from the Vest Fjord. But the cold-water mass from the Størfjordrenna extends into the ocean, and might prevent northward movements by the fish, thus making a "traffic jam." Whatever is the cause of the movement, two facts emerge: (1) the fish concentrate against the $2°$ C isotherm on the seabed; and (2) as the temperature boundary recedes eastward, the fish diffuse over the shelf.

The concentration of fish may thus be based on a form of kinesis, if the temperature barrier were sharp enough. As fish swim more slowly in cooler water, so they gather in patches. But this hypothesis does not explain why cod are not found at all in the really colder water. In the sea, temperature barriers are often not very sharp, the fish having to swim considerable distances before encountering a noticeable temperature change. But on the western Svalbard shelf in June, the temperature barrier can be fairly sharp — as much as $1.5°$ C in a mile (Lee, 1952). Fish are sensitive to a change in temperature of $0.03°$ C (Bull, 1936), and since 70-cm cod can cruise at 4.2 knots (Harden Jones, 1963) they could detect changes every 20 sec on a straight course. There is perhaps enough sensory information received by the fish to allow them to gather as the water gets colder. Such a minimal mechanism would allow the fish to gather in patches close to the cold-water boundary.

These studies suggest that the fish live naturally in the warmer water, that they might suffer some physiological disadvantage in penetrating cold water, but that there is probably a behavioral mechanism which returns them to the warmer water mass. The more general study in Chapter 3 suggests that the Arcto-Norwegian cod stock is restrained in area by a cold-water boundary throughout the Barents Sea. This is sensible because there is less food in a broad sense in the cold water, but paradoxically enough, that boundary — at the surface at least — is where food is produced (Marshall, 1957).

The Herring and Its Boundaries. One of the first pieces of research carried out by the Norwegians after the Second World War resulted in the discovery of herring offshore in the Norwegian Sea in winter, before they reached the traditional grounds close to the coast (Devold, 1952). In later years the work was extended to the whole Norwegian Sea and at other seasons than midwinter. From 21 May to 30 June 1954, sonar and temperature surveys were made (Fig. 88) by research vessels from Denmark, Iceland, and Norway (Tåning, Einarsson, and Eggvin, 1955). The temperature distribution conveniently describes a complex oceanic state. The warm Atlantic current moves between the Shetland Islands and the Faeroes along the Norwegian coast; in the northern Norwegian Sea, it breaks into the West Spitsbergen current and the North Cape current. The synoptic temperature distribution also describes a cold mass of water between Iceland, Greenland, and Jan Mayen. This is the polar front at which the herring used to gather in summertime, before the Atlanto-Scandian stock was reduced so far as to extinguish all fisheries on that stock. Up to 600 Russian drifters worked near the polar front where herring fed in the highly productive waters in which plankton was generated as the ice edge receded.

Off the east coast of Iceland, the warm Atlantic water in the Irminger current makes a sharp boundary with the cooler water, near which the herring sonar traces were found. To the north, extensive traces were found over a

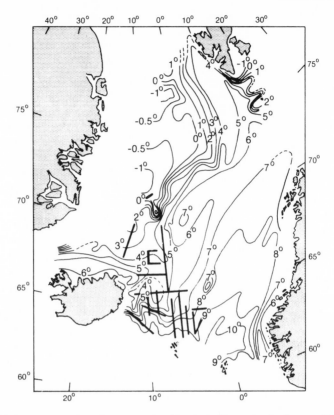

Figure 88. The concentration of herring at a boundary in the Norwegian Sea between 21 May and 30 June 1954. The bold black lines represent the presence of fish, as shown by sonar and echo-sounder records. The temperature observations were made at 20 m. Adapted from Tåning, Einarsson, and Eggvin, 1955.

wide area of cool water. There is no simple relation between herring and temperature, because the fish are found in dense patches at all temperatures from 2° C to 9° C, which is the total range of temperature in the Norwegian Sea during the period of survey. Yet the distribution of sonar traces bears some relation to the current structure as outlined by the temperature distribution. The East Icelandic current swings southward and southeasterly toward the Faeroe Islands; in it, the Atlanto-Scandian herring gathered in a few large shoals before they migrated in autumn toward the Faeroes. Later in the year, the sonar traces moved eastward toward the Norwegian coast and then appeared to "punch" their way through the warm-water barrier of the Atlantic extension. Devold (1966) showed that the fish cross from the Faeroe Islands to the Norwegian coast at a depth of 150 m in about seventy days. The distance is about 360 nautical miles, and they migrate no more than 5 nautical miles/day on a straight course. The invention of the hydraulic power block in California in the late 1950s allowed the purse seiners to move from sheltered waters on the Norwegian coast out to the open sea. They found the large shoals in the East Icelandic current in autumn in an inhospitable sea and reduced them to unprofitable quantities.

The whole migration of the Atlanto-Scandian herring has been described by Devold (1963) as a movement away from the coastal spawning grounds across the Norwegian Sea to the polar front, which itself is one of the sources of the East Icelandic current. The full circuit resembles that of the arctic cod to some degree, in that the polar front forms an edge to it which is not crossed and which would be unprofitable to cross. However, the problem is complex and a simple interpretation does not seem applicable because the nature of the migratory mechanism is not understood. In particular, the movement to the polar front and the movement beneath the Atlantic stream to the Norwegian coast remain mysterious.

Another boundary at which herring gather has been described by Steele (1961) in the eastern North Sea in spring. The deep water off the Norwegian coast is overlaid by the Baltic outflow, flowing north. The outflow of cold, somewhat fresh water at the surface forms a stable layer on top of the saltier, but warmer, North Sea water. Because the layer is stable there is little vertical turbulence, in sharp contrast to the North Sea itself in March and April. Consequently the production cycle starts much earlier in the Baltic outflow, where both chlorophyll content and quantity of zooplankton are high. The fish appear to live in the daytime below the Baltic outflow and migrate upward at night to feed on euphausiids and *Calanus*; the herring tend to stay on the western edge of the deep water off Norway, which coincides with the edge of the Baltic outflow. An apparently simple association of echo trace (or herring) and surface temperature is probably the consequence of a rather complex mechanism.

Although herring were caught predominantly in the open North Sea in summer, or in the tidally dominated southern North Sea and eastern English Channel in autumn and winter, the fish elsewhere were caught at boundaries. As all are based upon seasonal sectors of migration circuits (see Fig. 20, Chapter 2), such fisheries are regular in their appearance in time and space, as the fishermen well know.

Tuna and the Current Boundaries. The Gulf Stream and the Kuroshio are fast currents with sharp boundaries. The current is fast at the boundary because the latter is the edge of an oceanic gyre, and Stommel (1958) has shown that, in the northern hemisphere, currents are swifter on the western side of an ocean. Uda (1959) has found that the boundaries within the oceans are good places to fish because of the concentrations, particularly of sardine-like fish, that occur there; the Japanese fishermen call these edges "siome," which can be seen at the surface or even heard as one opposing surface rubs against another. The bluefin tuna are caught between January and March along the divergences at the western boundary of the Kuroshio off Kyushu, the southernmost island of Japan. Uda (1952) plotted the positions of catches from year to year and showed that in 1937 and 1940, when the axis of the Kuroshio was displaced southward, the effect was that the fish were concentrated more sharply at the current boundary and high catches were made.

Suda (1963) believed that one stock of albacore live in the North Pacific anticyclone and another one in the South Pacific anticyclone. In the north, large fish are caught in the North Equatorial current, and in the Kuroshio and Kuroshio extension small fish predominate in the catches. Otsu (1960) showed that fish tagged off California crossed the Pacific Ocean in less than a year. A more detailed examination of the catches shows that spawning albacore are taken in June and July in the North Equatorial current, and larvae are probably distributed across much of the ocean (Matsumoto, 1966). The adults are subsequently caught in winter off Indonesia, in the South China Sea, and in the Kuroshio itself. In the following summer immature fishes appear in the catches in the Kuroshio extension, probably from the spawning a year earlier in the North Equatorial current (Nakamura, 1969). In March and May of 1954 nuclear bombs were detonated at Bikini atoll in 165° E in the North Equatorial current; six months later, radioactive fish were found in the Kuroshio extension in 176° W, a movement of three thousand miles in six months. Similarly, radioactive fish appeared off eastern Australia, presumably via the South Equatorial current. Thus, the albacore make use of the whole North Pacific anticyclone in their migrations, and fishermen catch them where they are concentrated in the current structures. At the Equator, there are two main currents (see Fig. 92, below): the North Equatorial current westbound between 10° N and 18° N, and the South Equatorial current westbound between 2° N and 14° S. Between them in the Pacific runs an Equatorial countercurrent eastbound between 5° N and 10° N. Between the countercurrent and the South Equatorial current, there are zones of divergence and convergence. A zone of divergence is one where water rises from below and diverges in the surface layer; conversely, a zone of convergence is one where water sinks from surfaces that converge toward each other. Divergence and convergence are to some extent complementary aspects of the same phenomenon.

King and Hida (1957) found that the yellowfin tuna occurred in the zones of both convergence and divergence of the Pacific equatorial currents, and that the catches were between two and three times greater than in the main currents (Fig. 89). The zooplankton volumes are greater in these zones, but only marginally so, as compared with the South Equatorial current in the second half of the year, or with the Equatorial countercurrent in the first half. It might be thought that the yellowfin aggregate on the zooplankton, but they feed on rather larger animals — fishes the size of sardines or myctophids (Legendre, 1934) — which, in turn, feed on zooplankton. The tuna must aggregate transversely across the currents into the zones of divergence and convergence. The process of aggregation implies the existence of a reservoir from which the fish can move to concentrate, and the equatorial currents may well provide this. It is often thought that divergence generates more nutrients, which produce more zooplankton. If this was actually the case, the zone of convergence should be sterile, which it is not.

Murphy and Shomura (1955) have shown that the degree of aggregation of the yellowfin tuna depends on the temperature structure of the water. When the zone of divergence is narrow, the fish are tightly packed, but when the cool water is spread over a wide area, there is little aggregation.

The boundaries along the Equator are complex, and the mechanisms of fish aggregation there are not fully understood; neither is readily explained in terms of simple associations. There are two ways in which the mechanisms could be elucidated — a study of the causes of aggregation, and an examination of the means by which the tuna migrate. However, the current boundaries that limit the tuna distributions restrain the individuals probably in a hydrodynamic sense rather than a physiological one. But the scale of such a gen-

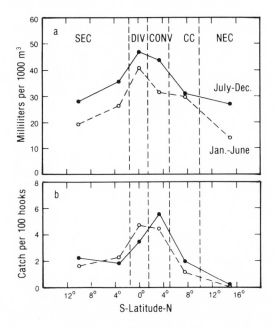

Figure 89. Catches of zooplankton (*a*) and of yellowfin tuna (*b*) in a transect across the Equator. SEC = South Equatorial current; DIV = zone of divergence; CONV = zone of convergence; CC = countercurrent; NEC = North Equatorial current. Adapted from King and Hida, 1957.

eralization is enormous — the breadth of the Pacific. On a smaller scale, the movement northward of albacore off the California coast has been related to particular temperatures (Laurs, Yuen, and Johnson, 1977). The real problem in any study of the fishes that make the fisheries is to establish the scale at which the critical events occur.

Upwelling. With the term "upwelling" used in a broad sense to include divergences and other oceanic processes, some of the areas of upwelling important to fisheries are shown in Figure 90. This is Townsend's (1935) chart of the positions of sperm whale (*Physeter catodon* Linnaeus) catches which were logged by whalers working from Nantucket, Massachusetts, between 1729 and 1919. From October to March there were four main areas of catching which have been identified as upwelling areas — in the Peru current along

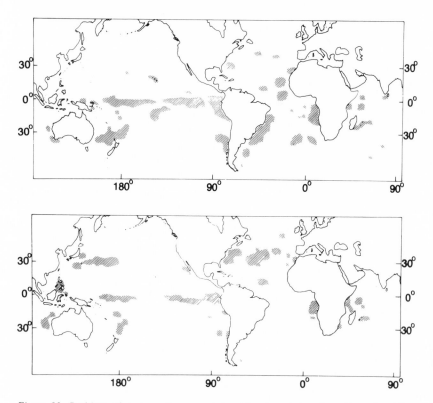

Figure 90. Positions of capture of sperm whales taken by whalers from Nantucket, Massachusetts, from 1761 to 1919. *Top:* October–March, the southern summer and northern winter. *Bottom:* April–September, the southern winter and northern summer. The major upwelling areas are shown off California, Peru, northwest Africa, and southwest Africa, together with the main equatorial system in the Pacific, the Line. From Townsend, 1935.

the western coast of South America, in the Benguela current along the western coast of South Africa, off the Cape Verde Islands, and along the Equator; there was another area off the southern coast of Somalia. From April to September, in addition to the Equator and the Benguela areas, there was another near the Kuroshio extension east of Japan. Many are areas of upwelling, but there are others which are not, like the Sargasso and the area just off Nantucket.* Townsend's picture is one of the simplest ways of representing the upwelling areas, presumably because the sperm whales aggregate on the fish stocks gathered there. Similar charts of guano islands and phosphatic deposits also indicate upwelling areas.

There are four main upwelling areas in the eastern boundary currents of the subtropical anticyclones in the Atlantic and Pacific: off Peru, South Africa, California, and northwest Africa. There are many minor upwellings within particular coastal configurations, for example in the Gulf of Panama and in the Gulf of Tehuantepec, in Mexico. In the Indian Ocean, there is a major upwelling in the Somali current during the Southwest monsoon (April to September) off Somalia and off the southern Arabian peninsula. There are less important ones off the coasts of India, Burma, Java, and Vietnam. One of the most extensive areas of upwelling is found in the equatorial system.

In an upwelling area, the wind blows alongshore toward the Equator (west of a continent), and the water flows to the right of the wind in the northern hemisphere, or to the left of it in the southern hemisphere, so water drifts offshore. The movement at right angles to the wind is due to Coriolis force, which expresses the effect of the earth's rotation. Figure 91 shows the upwelling mechanism in the Benguela current; water is drawn from up to 200 m from the shoreline to a dynamic boundary, or "roller bearing," about 50 km offshore. Beneath the upwelling system a deeper countercurrent flows away from the Equator, whereas the broad surface current beyond the dynamic boundary flows toward the Equator.

The central problem in a study of upwelling is how the huge fish stocks are generated, how their numbers are maintained, and how the fish remain in the same position. In the deep tropical ocean far from the upwelling zones, the water is very clear and the plankton is very thin. The appearance of cold water at the surface generates dense plankton outbursts — either because the water is cold, or because it is rich in nutrients, or both.

A stable cycle (× 10 in amplitude) of plankton production in deep tropical oceans far from the upwelling areas may be contrasted with what has been described as an unstable, or unbalanced, cycle (× 100 or more in amplitude) in temperate waters (Cushing, 1959b). The difference between the two cycles is that the temperate one does not produce in winter, whereas the tropical

---

*These two areas may have been searched much more systematically than others, because the whalers had to cross both areas when leaving and returning to Nantucket.

cycle is continuous; so, when the temperate one starts again in spring, the plant and animal components are out of phase with each other. The plants grow to very high densities before the animal populations, which control them by grazing, have a chance to increase in numbers. The productive cycle in an upwelling area resembles the temperate cycle in that the whole cycle starts from very low initial numbers; the appearance of cold water at the surface is analogous to the temperate spring. Hence, the cold water and the initially sparse plankton are sufficient to account for the subsequent dense outburst. At this stage, the accompanying rich nutrients are perhaps irrelevant; but at a later stage they may be important. There is some evidence in temperate waters that the spring outburst is controlled less by lack of nutrients than by grazing (Cushing and Nicholson, 1963). By keeping zooplankton-free water in a large plastic bag hung at the ocean surface, McAllister, Parsons, and Strickland (1960) have shown that the algal quantities increased by many times. Hence, the low algal quantities in the deep ocean must be controlled at that low level

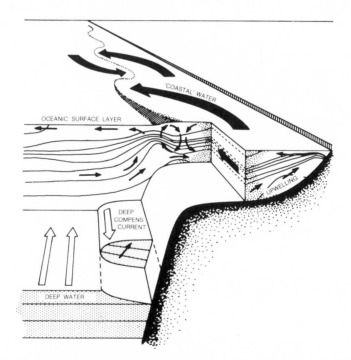

Figure 91. The mechanism of upwelling in the Benguela current off the southwestern coast of Africa. The longshore breeze blows the surface waters offshore under Coriolis force toward the dynamic boundary (or "roller bearing") between the coastal upwelling system and the main oceanic structure. The main water mass is drifted northward toward the Equator in the eastern boundary current at the surface. Beneath the upwelling system, a countercurrent (or deep compensatory current) flows poleward. Adapted from Hart and Currie, 1960.

by the grazing animals. In temperate waters, however, nutrients decline after the spring outburst, being absorbed principally as animal flesh and possibly limiting subsequent production. In upwelling areas, there is a continuous presence of nutrients after the first outburst. High production thus appears to be maintained continuously at the point of upwelling, the subsequent cycles supposedly decaying at a distance as the upwelled water drifts out into the deep ocean.

Fish can either aggregate from elsewhere to the upwelling region or the year classes can build up at the point of upwelling as a result of the larvae drifting with the plankton. Nothing is known decisively of the migratory habits of fish in the upwelling areas, but it seems unlikely that they aggregate from elsewhere, because upwelling is a more or less continuous process, at least in the Peru and Benguela currents. Hence, food for larval fish is permanently available. Large recruit classes have been built up at very high stock levels. Knowledge of the migratory cycle is still needed, but one would guess that the baby fish grow up, not close inshore like North Sea juvenile herring, but offshore where the plankton is very dense. Hence, when food is especially abundant, the recruit classes are strong.

The most extensive upwelling region is the broad area about the Equator where Nantucket sailors caught sperm whales long ago. The current structure of the eastern tropical Pacific is displayed in Figure 92. The two subtropical anticyclones flow westerly as the North Equatorial current and the South Equatorial current; between them flows the Equatorial countercurrent easterly

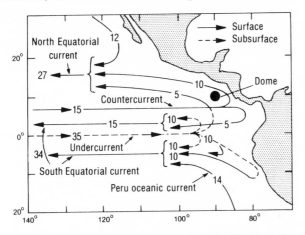

Figure 92. The current structure of the eastern tropical Pacific. The North Equatorial current flows westerly as part of the northern subtropical anticyclone; the South Equatorial current flows westerly as part of the southern subtropical anticyclone and its northern edge lies just north of the Equator. Between the two anticyclonic currents the Equatorial countercurrent flows easterly toward Costa Rica. At the Equator in a depth of 150 m the Equatorial undercurrent flows easterly from midocean toward the Galapagos Islands. Adapted from Wyrtki, 1966.

in latitude 8° N. The latter splits at the continental shelf, turning north and south into the two equatorial currents. At the Equator at a depth of 150 m flows the Equatorial undercurrent, which is narrow and swift; high primary production occurs above it in the eastern Pacific and about the Galapagos Islands (Houvenaghel, 1978). Minor and sporadic upwellings occur in the divergences of the equatorial currents where the sperm whales gathered.

Two most remarkable distributions are shown in Figure 93*a* and 93*b*: the phosphate-phosphorus distribution in the top 100 m, and the zooplankton distribution in the top 200 m, in the Pacific Ocean. Many structures are illustrated in the distribution of zooplankton — the major upwellings off California and Peru, the broad divergences of the South Equatorial current and the empty countercurrent just north of it, the region of divergences in 40° N on

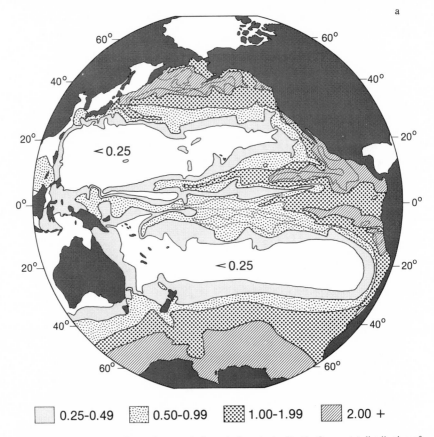

a

□ 0.25-0.49   ▓ 0.50-0.99   ▓ 1.00-1.99   ▨ 2.00 +

Figure 93. Distribution of phosphates and of zooplankton in the Pacific Ocean: (*a*) distribution of $PO_4 - P$ in a depth of 100 m at five levels in μg atoms/l; (*b*) distribution of zooplankton in ml displacement volume from about 200 m to the surface, with some observations from 300 m and others from 100 m. Distribution of phosphates shows the high production of the Alaska gyre, the

the southern edge of the Alaska gyre, the rich area of the Oyashio current in the northwest Pacific and that in the antarctic. Phosphorus is taken up by animals, and we might therefore expect zooplankton and the phosphorus absorbed to be correlated. Figure 93 shows an association between zooplankton and the phosphorus remaining in the sea, which can be explained quite readily if the phosphorus in the sea is primarily composed of phosphorus regenerated by the animals. It is possible that the production cycle in temperate waters is often restrained when the transfer of nutrients to animal flesh is greater than the regeneration, which is an indirect way of describing the dominant effect of grazing in the temperate cycle; by the implication of Figure 93, the same is true of the world ocean.

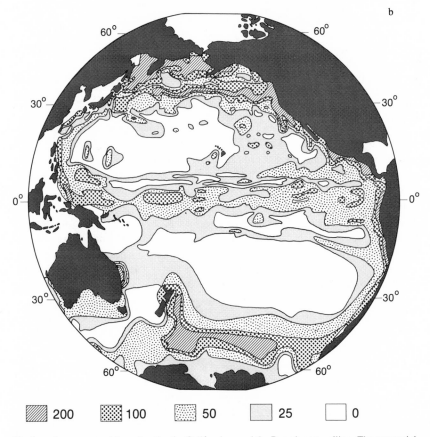

Northwestern gyre, and the antarctic, the Californian, and the Peruvian upwelling. The equatorial region is also productive; the two equatorial currents are shown in the distribution of phosphates. The same points are shown in the distribution of zooplankton; the Equatorial countercurrent can be seen. Adapted from Reid, 1962.

One of the most productive fisheries in the world is that for the anchoveta in the Peru current off Peru and northern Chile (Jordan and de Vildoso, 1965). With increased exploitation in the late 1950s, catches reached about one million tons in 1959 and increased steadily to a peak of 12 million tons in 1971. However, the year classes of 1971, 1972, and 1973 were reduced by about an order of magnitude, and since then annual catches have been about 3 million tons or less (Santander, 1976). The cold Peru current from the subantarctic region sweeps north along the South American coast, and wells up against the continental shelf between Peru and the Galapagos Islands. To give a description in more detail, there are two currents off Peru, the Peru coastal current within 50 km of the shore, which is the region of coastal upwelling, and the Peru oceanic current, the region of divergences offshore of 50 km. Between them flows a subsurface countercurrent, and below the coastal one flows an undercurrent. South of Pisco (14° S), cool water upwells from the lower levels of the coastal current, but to the north, the process is intense enough in winter to draw cool water from the countercurrent (International Decade of Oceanic Exploration, 1976).

Where the cold water reaches the surface, the plankton becomes dense and the fish gather. Like the Benguela, the Peruvian upwelling area is most productive. The primary production of carbon is high; the earlier observations averaged 0.67 g C/m²/day (Cushing, 1971a), but later, more extensive, ones averaged 1 g C/m²/day (Guillen, 1976), a range of 0.8–5.0 g C/m²/day. The total primary production amounts to about 110 million tons C/yr; secondary production has been estimated to be about one-tenth of that quantity. Anchoveta live in the upper part of the thermocline, with horse mackerel below them; hake and rosefish (*Sebastes marinus* [Linnaeus]) live near the seabed. The anchoveta spawn between Punta Aguja and San Juan in late winter and early summer when the Ekman transport is most intense and where the primary production is highest (Cushing, 1971a). The anchoveta eggs and zooplankton "exclude" each other in space (Flores, 1967; Guillen, in Flores, 1967), which suggests that the spawning fish select the phytoplankton patches so that the larvae may feed on the subsequent larval zooplankton.

Every few years off the northern coast of Peru the El Niño current appears, called "El Niño" because it comes at Christmas, the season of the Christ child. It is a warm salty layer extending south over the Peru current and accompanied by heavy rain; Bjerknes (1961) associated the phenomenon with atmospheric changes that cause a transequatorial flow from the Equatorial countercurrent. During a period of intense southeast trades, water is assumed to accumulate in the western Pacific. When the trades relax, the water should flow back into the eastern Pacific in the Equatorial countercurrent, but perhaps it does so also in the Equatorial undercurrent and in the South Equatorial countercurrent. The occurrences of El Niño have been correlated with anomalies in atmospheric pressure difference between Darwin, in Australia,

and Easter Island (Quinn, 1974; Wyrtki, 1973); El Niño occurs when the
difference is low, when the trades have relaxed. The phenomenon took place
in pronounced form in 1891, 1925, 1941, 1953, 1957–58, 1965–66, and
1972–73. The last event proved disastrous to the anchoveta fishery because
the year classes that hatched in those years (and in 1971) were sharply reduced. As the current moves south, it destroys the cold-water plankton. The
plankton rots in the water, and off Callao a sulfurous emanation is produced,
known as the "Callao Painter," which stains the white paint of ships, tar-
nishes silver in coastal homes, and may even kill large numbers of fish and
birds. A major fisheries problem in Peru is to determine whether there are
enough anchoveta for both the fishing industry and the guano industry. The
guano industry depends on the populations of the three species of birds living
on the Peruvian islands — cormorants (*Phalacrocorax bougainvillei* Lesson),
boobies (*Sula variegata* Tschudi), and pelicans (*Pelecanus occidentalis the-
gus* Linnaeus). In 1958, when El Niño came in, the bird populations were
sharply reduced. By 1962, however, they had fully recovered.

During the 1960s the anchoveta stock, feeding partly on phytoplankton,
supplied enough fish annually for 18 million birds as well as for the catch of 7
million tons taken by the fishermen. The recovery in the stock of the guano-
producing birds took place at a time when the fishery was increasing sharply.
But in the long term the number of birds has declined from 20 million in 1955
to 16 million in the period 1961–65, 4 million in 1966–70, and about 2 million
in 1973. Although the population recovered from the 1957–58 El Niño, it did
not from that of 1965–66 (Santander, 1976). Thus the guano industry has
suffered considerably from competition with the fishermen.

The events of 1972–73 were studied extensively by the Instituto del Mar in
Lima. El Niño appeared in February and March 1972 from the region of the
Gulf of Panama. The winds were weak, with heavy rainfall at sea. Primary
production was reduced to a third (Guillen, 1976) and a large number of
"indicator" species appeared from Ecuador, from California, and from the
subtropical ocean offshore; however, the most remarkable invader was the
milkfish (*Chanos chanos* Forskael), from the western Pacific, perhaps borne
by the Equatorial countercurrent (Vildoso, 1976). The pelagic crab *Euphylax
dovii* Stimpson appeared from tropical waters in the second half of 1971 off the
southern coast of Peru, where the water was warm, an event regarded by
Wooster and Guillen (1974) as an early signal of El Niño. The failure in the
fishery started with the reduced recruitment in 1971, before El Niño appeared,
but the reduction might have been associated with delayed maturation. Subsequent year classes were low, and the stock has not yet recovered.

The same type of problem connecting guano production with a fishery is
found in the Benguela current, where there are three sardine fisheries — at St.
Helena Bay in South Africa, Walvis Bay in Namibia, and Baia dos Tigres in
Angola (Davies, 1957). Between 200 thousand and 300 thousand tons of

South African pilchard and 80 thousand tons of maasbanker were caught each year (Davies, 1957). Figure 91, above, shows the mechanism of upwelling in the Benguela current (Hart and Currie, 1960). Where the cold water reaches the surface, plankton becomes very dense; in Walvis Bay, for example, it can become so dense as to rot on the shoreline in heaps. To produce 1.6 thousand tons of guano annually in this area (Fig. 94), the guano-producing penguins (*Spheniscus demersus* Linnaeus), gannets (*Sula bassana* Linnaeus) and cormorants consume 43 thousand tons of pilchard and 7 thousand tons of maasbanker (Davies, 1958). The fisheries annually account for an additional 200–400 thousand tons of pilchard and maasbanker. The proportion of the number of fish taken by birds is not only interesting in itself, but constitutes one of the first steps in a study of the cause of natural mortality.

The upwelling areas of the subtropical ocean are of dominant importance to the fisheries. Catches of sardine-like fishes amount to 15–20 million tons, and there remain areas in the Indian Ocean and Indonesian areas which are unexploited. The larger fisheries have been found in the four main upwellings in the eastern boundary currents. Much of the oceanic tuna fisheries are concentrated in the divergences off the main upwelling areas, along the Equator, and on the poleward edges of the anticyclones. An important point about the major

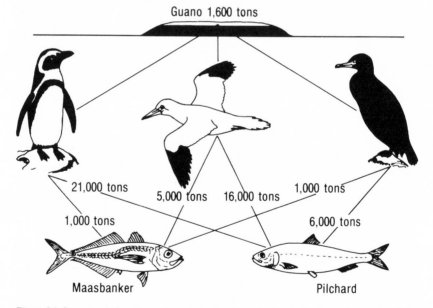

Guano 1,600 tons

21,000 tons

5,000 tons   16,000 tons

1,000 tons

1,000 tons

6,000 tons

Maasbanker   Pilchard

Figure 94. Some natural enemies of the pilchard and maasbanker in the Benguela current off the southwestern coast of Africa. The guano-producing birds are a species of penguin, a gannet, and a cormorant. Such studies could lead to an independent estimate of natural mortality. Adapted from Davies, 1958.

areas is that the structure of current and countercurrent establishes a biological unit. For example, the California hake spawns off southern California, yet the adults are caught off Oregon and Washington State; hence the larvae or juveniles must drift north in the countercurrent or in an inshore current. If that were true, the adults would move south in the main eastern boundary current beyond the dynamic boundary. The zooplankton populations probably drift through such an area in 40 days or so, yet the permanent fish populations exploit them.

The Response to Climate

It has long been known that some fish stocks appear and disappear in long periods of time, as do some herring stocks; fishermen know that such events are often related to long-term climatic changes. A long time series of cod catches in the Vest Fjord, in northern Norway, was related by Ottestad (1969) to differences in the widths of pine-tree rings in the area. Four periods were detected in the widths of the rings, and their components were added for each year. The resulting curve was fitted to the annual catches (lagged by seven years to allow for the effects of recruitment on the annual catches) by least squares for a period of 85 years, as shown in Figure 95. A similar analysis was made by Zupanovitch (1968) on the Adriatic sardine. Hence the recruitments to the cod and sardine stocks are modified by factors common to fish and to tree rings.

Templeman (1965) examined the year class strengths of ten stocks of cod and herring in the North Atlantic after the first two decades of the century and found that the outstanding year classes were common to all stocks throughout the ocean. The famous 1904 year class with which the recent Norwegian herring period started was also strong for the Arcto-Norwegian cod and the Icelandic haddock. In 1950 nine out of the ten stocks were outstanding. Hence the factors that affect the recruitments to the fish stocks are pervasive on an

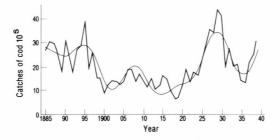

Figure 95. Catches of cod in the Vest Fjord, in northern Norway, lagged by seven years to indicate differences in recruitment. The line fitted by least squares was constructed from the sum of four periods detected in the widths of pine-tree rings in the same area. Adapted from Ottestad, 1969.

oceanic scale. Further, the differences from year to year indicated in the work of Ottestad (1969) and Zupanovitch (1968) may be common to the broad ocean.

Templeman (1972) selected quantitative material on year class strength of gadoid stocks and scored them on an arbitrary scale during a period of thirty years or so, from 1941 to 1970. Figure 96 shows the annual scores added each

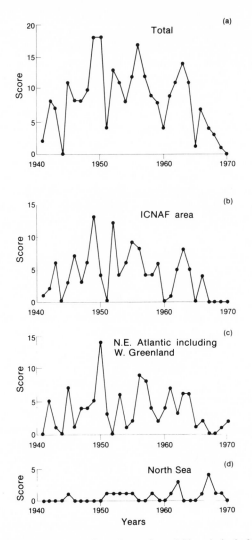

Figure 96. The sums of annual scores of recruitment for gadoid stocks in the North Atlantic from 1941 to 1970. Adapted from Templeman, 1972.

year in the Northeast Atlantic, the Northwest Atlantic, the North Sea, and for the whole North Atlantic. Recruitment throughout the ocean reached a peak in 1949–50, after which it declined; in the North Sea, there was an increase in the 1960s. As will be shown below, the period 1940–70 corresponds with the period of cooling that has occurred since 1945–50, ending a period of warming that began in 1880. A point of minor interest is that a good year class in the Northwest Atlantic is often succeeded by one in the Northeast.

The Nature of the Link between Climatic Factors and Recruitment. In temperate waters, the production cycle varies in amplitude, spread, and time of onset (Colebrook, 1965). Differences in time of onset, in amplitude, and spread are governed by annual differences in wind strength, wind direction, and in solar radiation. Hence changes in climatic factors from year to year, or from decade to decade, can affect the time of onset directly. In temperate waters the peak date of spring spawning is more or less constant in that the standard error of its mean is less than a week (Cushing, 1969a). Then the larvae drift away from their fixed spawning ground toward their fixed nursery ground into the spring outburst of plankton, as shown by Cushing (1967) for the herring populations of the Northeast Atlantic. The development of the production cycle is governed by the ratio of the depth of the euphotic layer to the depth of mixing. As spring advances, the euphotic layer deepens with the rising angle of the sun and the depth of mixing becomes shallower as the wind slackens. During the larval drift physical factors such as wind strength, wind direction, and solar radiation can affect the production cycle and hence the production of larval food directly. Hence, if recruitment is to be determined to some degree by the physical factors associated with climatic change, the critical events probably occur during the larval drift. The argument presented in Chapter 7 reaches a similar conclusion from another point of view.

The Recent Warm Period. Between 1880 and 1969, mean surface air temperature anomalies in the northern hemisphere reached a peak in 1945 of 0.4° C (Cushing and Dickson, 1976). The recent period of cooling has been associated with weakened westerlies over the British Isles and meandering Rossby waves in the northern hemisphere. The preceding period of warming was marked by two forms of invasion in the sea. The first was the appearance of subtropical animals off California, in the Bay of Fundy, off the western British Isles, the Faeroes, and the southern coasts of Iceland; such animals (albacore, *Velella, Ianthina, Physophora*, etc.) were conspicuous and were recorded particularly in the decade of 1925–35. The second invasion was that of boreal animals which appeared during the same period off West Greenland, north and east of Iceland, off the Murmansk coast, off Spitsbergen, and on the Svalbard shelf. Each invader is a stray from the normal migration circuit, and the northerly spread of migrants suggests a northerly shift in the migration circuit, perhaps associated with an intensified circulation.

The most remarkable invasion during the period of warming was the rise and fall of the West Greenland cod fishery. Cod were taken off West Greenland in the nineteenth century, but then disappeared; in 1912, on the Tjalfe expedition to the offshore banks of West Greenland, a few cod were found. A new stock established itself with the year classes 1917, 1922, 1926, 1934, and 1936, probably from a larval drift from the cod spawning ground near the Westman Islands off southern Iceland. During the 1920s and 1930s the fish appeared farther and farther north off the coast of West Greenland. The strongest year classes appeared in 1945 and 1949, and international catches in the 1950s and early 1960s amounted to as much as 450,000 tons/year. The last good year class arose in 1963 and the last significant one in 1968; now the stock is so low that catches off Greenland are banned. The period of the West Greenland cod fishery is roughly that of the recent period of warming, with cooling supervening after 1945.

Cod tagged off West Greenland (Hansen, Jensen, and Tåning, 1935) were recaptured as mature fish on the Icelandic spawning ground; indeed, between one-fourth and one-third of the Icelandic spawning stock was sustained by Greenland migrants. The life of this fishery represented the most dramatic example of conspicuous invaders during the recent period of warming. Dickson et al. (1975) have shown that between 1930–39 and 1956–65 the Iceland low shifted south of Greenland, which generated easterly winds across the Denmark Strait. Such a breeze might have assisted the larval drift from Iceland to Cape Farewell in the Irminger current and the East Greenland current. By 1966–70 the Greenland high shifted, and northerly winds blew in spring across the Denmark Strait, a change which might have prevented the larvae from drifting from Iceland to Greenland. Such is the form of mechanism which might have stimulated and stopped the most dramatic migratory invasion during the recent period of warming.

In more general terms, as the Greenland fishery declined, stocks of gadoids in the North Sea increased. The haddock year class of 1962 was twenty-five times larger than the preceding average; equally strong recruitments appeared in 1967 and 1974. Big whiting year classes appeared in the same years and good cod year classes came in 1963, 1964, 1965, 1969, and 1974. Stocks of gadoids in the North Sea have increased by nearly an order of magnitude during the 1960s, and yields have increased by a factor of three. In earlier centuries such a change from northern latitudes to the North Sea might have been interpreted as a southerly migration. No such migration took place, but a decline in the north was succeeded by an increase in the south. Thus two processes were mingled, the migration of conspicuous animals and cod larvae into northern waters, perhaps in a period of extended circulation, and the rise and later fall of populations *in situ* as the mechanisms of recruitment were modified.

Long-Term Changes. During the recent warm period that occurred between 1880 and 1970, the Norwegian herring fishery flourished, specifically from 1904 to 1967. Since the Middle Ages this fishery has alternated roughly with the Swedish herring fishery, and the latter is characteristic of a cooler period. Figure 97 shows the presence or absence of the two fisheries, which have alternated with a period of about a century since the early fifteenth century. Ljungman (1882) fitted a period of 110 years to such observations of the Swedish catches, but not very successfully, as shown by the line of crosses; however, peak catches, shown by the circles, occurred at the center of the periods. The figure also shows that periodic changes in the Japanese sardine fishery corresponded with those of the Norwegian or Atlanto-Scandian herring, but that changes in the Hokkaido herring fishery were independent of the others. Zupanovitch (1968) has shown that catches of sardines off California and Spain and in the Adriatic reached a peak in the 1930s, during the Norwegian herring period. This period is that of the West Greenland cod fishery and indeed that of high cod production in northern waters, as shown in Figure 96.

Changes in the whole marine ecosystem also occurred during the recent period of warming. In the western English Channel between 1925 and 1935 a number of changes took place, which then reversed between 1966 and 1976. Pilchard eggs were noticed first in 1926, but they became commoner and reached high numbers in 1935, after perhaps three generations. They replaced a herring population, of which the last good year class was hatched in 1926 and entered the fishery in 1930–31; the fishery finished in 1938, as the older year classes became extinct. In the autumn and winter of 1930–31, macro-

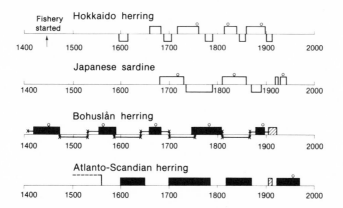

Figure 97. The presence or absence of the Swedish (Bohuslån) and Norwegian (Atlanto-Scandian) herring fisheries and the periods of high and very low catches in the Japanese sardine and Hokkaido herring fisheries since the fifteenth century. The periods indicated by lines marked with crosses were estimated by Ljungman (1882). The circles at the centers of the periods indicate peak catches. Hatched areas indicate periods of low abundance. Adapted from Cushing and Dickson, 1976.

226   FISHERIES BIOLOGY

plankton was reduced by a factor of four and the winter phosphorus maximum
by one-third. The numbers of larvae of spring spawners declined sharply in
1931 and those of summer spawners declined in 1935 (Cushing, 1961). They
increased again in 1966, as did the spring spawners in 1970; macroplankton
and winter phosphorus increased in 1970. The numbers of pilchard eggs were
reduced during the early 1970s (Russell, 1973), and by 1977 herring were
once more caught in the region. These most profound changes in a marine
ecosystem have been named the Russell cycle, after their discoverer (Fig. 98).

Such an ecosystem must comprise a large number of populations, perhaps
fifty or a hundred, but there must have been considerable changes in structure.
A system of herring, macroplankton, high winter phosphorus, and rich fish
larval numbers was replaced in the 1930s by one of pilchards, no macroplankton, low winter phosphorus, and poor numbers of larval fish. The numbers of elasmobranchs and of benthic animals also changed dramatically
during the period. The Russell cycle started in that decade when the conspicuous invaders entered northern waters, most noticeably when the cod became
established at West Greenland, and it ended in the decade when the West
Greenland cod fishery collapsed. Thus the ecosystem responded to the period
of warming much as did the fish populations. One of the remarkable points
about the appearance and disappearance of fish populations is that recruitment

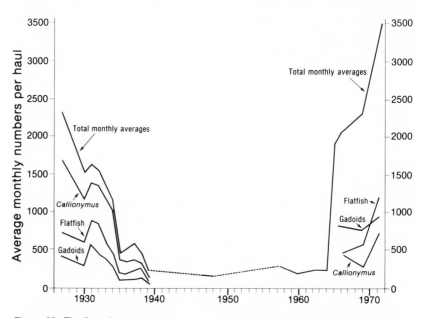

Figure 98. The Russell cycle: the changes in spring (gadoids and flatfish) and summer (*Cal-
lionymus*) spawned fish larvae, and in other components of the ecosystem which started in the
decade 1926–36 and which reversed in the decade 1966–76. Adapted from Russell, 1973.

must have increased or decreased by many orders of magnitude. There is a sense in which the fish populations and the English Channel ecosystem appear to rectify the periodicity of low amplitude as shown by the small changes in sea temperature in the northern hemisphere since 1880. But, also, the ecosystem responds to the same factors that modify the single population. This is not very surprising, as such events might be expected to modify the stock-recruitment relationships of populations other than fishes in the sea.

More remarkable is the change of structure observed in the ecosystem at the start and end of the Russell cycle. The recruitment to a single population may decline sharply under what we may loosely term climatic change. But that population is replaced within two or three generations, more rapidly than the well-documented replacement of sardine by anchovy off California (Ahlstrom, 1966). Most remarkable, the replacement is repeated at about the same time at all trophic levels. It is as if each population was understudied by one at a lower abundance ready to take over; indeed, there might be a reserve cast waiting in the wings.

Some Possible Mechanisms. Figure 67 in Chapter 7 shows detailed time series of recruitment for a number of stocks of different groups of fishes. One generalization is possible — that stocks of herring-like fishes show upward or downward trends with time and that stocks of cod-like fishes remain at more or less the same level in time. In other words, the herring-like fishes respond to environmental changes continuously, but the cod-like ones tend to be insulated from them. The argument can be pressed a little further in the long-term changes: the herring fisheries appear and disappear through the centuries whereas the cod fisheries (at least in the Vest Fjord) do not, although the stocks may vary in time. The difference between the two groups may reside in their respective stock-recruitment curves; that of the herring is lightly convex, whereas that of the cod is dome-shaped. Hence the cod stocks can respond quickly, whereas those of herring can only do so slowly. This property of resilience (Holling, 1973; Cushing, 1971b) may account for the responses of fish stocks to environmental change.

A more interesting point is how the changes in recruitment of many orders of magnitude might occur. Figure 99 shows a stock-recruitment curve and a bisector (which may be a misleading concept in a multi-age stock, but is used here for convenience because an analogue must exist in such a stock). At (a), in low stock, year classes vary, as shown in the vertical bar, and if $R > P$ the stock will stabilize itself. But if $R < P$, stock will be reduced; if the condition persists, stock must decline considerably — to the origin on the figure. But, in reality, the stock may only decline toward a lower level. At (b), in high stock, the stock can only stabilize itself if $R < P$, and if $R > P$ persistently, it will increase. Such a mechanism might account for the rises and collapses of recruitment of many orders of magnitude and for the rectificatory character of

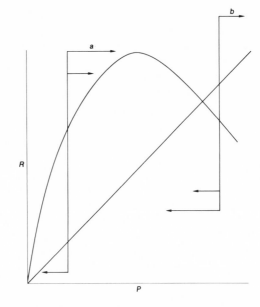

Figure 99. Diagram of a stock-recruitment curve with bisector (*R* = recruitment in numbers; *P* = stock in weight). At (*a*) the variability of recruitment is great enough to extend below the bisector, and hence persistently low year classes would lead to collapse. At (*b*) the converse effect is shown at high stock, which would lead to rapid augmentation of the stock. Adapted from Cushing, 1969a.

the long-term trends in herring-like fisheries and in ecosystems as in the Russell cycle. Such a mechanism implies that all populations in the ecosystem stabilize themselves much as fish stocks do. But it does not account for the reserve cast that may wait in the wings during any long-term period, as described above.

Such a mechanism may account for the changes of many orders of magnitude at the start and end of a "climatic" period. The question arises whether the pronounced variability of recruitment from year to year can be described in similar terms. The production cycle varies in time of onset, amplitude, and spread, yet in temperate waters spring spawning fish spawn at fixed seasons, as indicated above. Hence the production of larval food, which must vary as does the production cycle, is matched or mismatched to the production of the larvae. The principle is illustrated in Figure 100 at low stock and at high stock. Each half of the diagram consists of three parts, the production of eggs, of larvae, and of larval food. Each part is represented by a production curve in time, and above that of larvae and larval food is drawn an error bar. There is none above the production of eggs because the standard deviation of the peak date of spawning is low. The error in larval production is asymmetrical because production is an inverse power function of temperature; that is, at high temperature it is somewhat accelerated, but in cool water it is much delayed. But the error in the production in larval food is symmetrical and considerable. The overlap in time between the production of larvae and that of their food indicates the degree of match or mismatch. Such a mechanism has a virtue in that the three physical factors which might be associated with climatic variation (wind strength and direction and solar radiation) express themselves in the variability of the production of larval food. Another facet of the diagram emerges when high and low stock are compared. At high stock the overlap between the two distributions is greater, which may account for the observed higher variability of recruitment at low stock.

Figure 100 illustrates the match/mismatch hypothesis. It is now possible to model the production cycle (Steele, 1974) with as few as eighteen parameters, but a further step is needed. The model should be driven by the three physical parameters — wind strength, wind direction, and solar radiation — so that differences in them from year to year modify the time of onset, amplitude, and spread of the modeled cycle. Then it might become possible to estimate the degree of match or mismatch from year to year in the spring and to test the

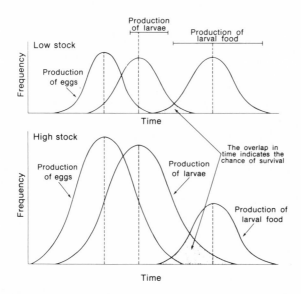

Figure 100. An illustration of the match/mismatch hypothesis. In temperate waters, eggs are laid at a fixed season, but the production of larvae is an inverse power function of temperature, and the production of larval food is highly variable. The overlap between the production of larvae and that of their food indicates the degree of match or mismatch of larval production to that of their food. Recruitment may be linked to this degree of overlap. At high stock, as compared with low stock, the increased overlap may show why the variability of recruitment may be less at high stock than at low stock. Adapted from Cushing and Dickson, 1976.

possibility that the major part of the variation in year class strength may be accounted for in the match or mismatch of larval production to that of larval food. More ambitiously, a construction of this nature might be used to trace the trends of year class strength with time, such as are illustrated in Figure 67.

The Institute of Marine Environmental Research, in Plymouth, England, has for many years operated a network of plankton recorder routes across the North Sea and the North Atlantic. The plankton recorder, a robust plankton net towed behind merchant ships at high speed, collects the material continuously on rolls of nylon netting. A summary of records from 1949–69 from the North Sea and North Atlantic expressed in standard measure (i.e., a unit of one standard deviation) shows that total copepoda and zooplankton biomass have declined when the time of onset of the spring bloom has been delayed by nearly three weeks (Glover, Robinson, and Colebrook, 1972). The phytoplankton is estimated as green color on the nets, which may not indicate quantities particularly well but which does display differences from year to year quite adequately. Such events indicate the type of trend that might be expected under the match/mismatch hypothesis, trends are often indicated by changes in average sea surface temperature from year to year.

The Generations of the Population in the Sea

The study of fish populations in space and in time is very rewarding because much is revealed about the nature of their response to the environment in which they live. The oceanic boundaries were examined originally as areas where fish gather and where fishermen might do also. But they are also the edges of the migratory circuits, the frontiers which individuals of unit stocks do not cross save as strays. Hydrodynamic regions such as the North Sea, the Alaska gyre, or the Gulf of Thailand are physically identifiable but are flushed in a relatively short time. Yet in such regions stocks retain their identity and remain for very long periods of time. Even in such temporary and variable structures as the upwelling areas, the fish stocks and indeed the whole ecosystem establish an unexpected unity. To understand such mysteries we need to understand the physiological mechanisms which govern migration and the structures which ensure that a stock retains its genetic identity.

The temporal changes in a fish population are expressed as recruitments in time series which may trend upward or downward or remain steady, yet such series may start or end abruptly. Physical factors, such as wind strength, wind direction, and solar radiation are modified, perhaps each differently as the climate slowly changes. Population control and the generation of year classes may be different facets of a common process, as described briefly in Chapter 7. It is a continuous adaptation by which the population estimates the quantity of food available. Indeed, one might imagine that the recruiting year class has explored and estimated this quantity. The annual variability may express the differences in food availability, which changes in time and perhaps trends upward or downward. The lack of trend in gadoid stocks in time may indicate a strong control in the population. In general, if recruitment has an exploratory function the stock has a conservative one, as it represents the average of a number of assessments made by the recruiting year classes.

Fish stocks are becoming well known, perhaps the best known of all large wild populations. They maintain large numbers across broad regions and preserve their reproductive isolation in recognizable provinces of the world ocean. To understand how they preserve the evolutionary urge, we need to establish the physiology of migration, the genetics of isolation, and the ecology of stabilization. The needs of management become ever more complex and demanding. The broader the base of fisheries biology, the greater the chance of satisfying those needs.

Summary

The nature of some aggregations at oceanic boundaries has been examined under a number of conditions. The concentration of cod against the 2° C isotherm on the seabed is described in some detail. Such mechanisms are at the base of the cod's migratory pattern, and the 2° C isotherm forms a reliable edge to the stock area of the arctic cod in the Barents Sea. The Norwegian herring are bounded in much the same way by the polar front between Iceland and Jan Mayen, and both boundaries are also highly productive.

For tuna and herring, there are simple associations of fish and temperature, which probably are indicators of more complex mechanisms. The differences in position of the bluefin fishery in the Kuroshio current vary with the displacement of that current by the cold Oyashio current and is thus a hydrodynamic displacement. It is likely that the aggregation of yellowfin tuna at the Equator is one of these complex mechanisms. That causing the aggregation of herring on the edge of the Baltic outflow in early spring has been discussed to some extent. The fisheries biologist is interested not only in the mechanisms themselves, but in the underlying reasons for the dependence of a stock like the North Sea herring on the edge of the Baltic outflow each year as a part of its regular mode of life. The mechanisms and their relationships to the stocks, when understood, will provide the biological bases of stock unities.

The examination of upwelling areas provides many opportunities for speculation. In such areas, the structure of productive cycles should be closely examined, together with the migratory pattern of the dominant fishes. The physical mechanism of upwelling provides at one and the same time high production and a current structure that allows a migration circuit to become established. The equatorial complex of current and countercurrent provides the most extensive spawning ground for tuna from one end of the Pacific Ocean to the other. Some progress has already been made in analyzing causes of natural mortality for the South African pilchard. Further, the anchoveta fishery off Peru may provide good opportunities for studying the problem of stock and recruitment because of the periodical extensive stock changes associated with El Niño. The recent changes in that anchoveta fishery are most complex from both the physical and biological points of view, and neither is well understood.

Due to the vastness of the oceans, extensive surveys have been more or less crude, and have thus far yielded little more than the rough associations described above. In each of the associations, the fisheries biologist will find complex mechanisms, as in the study of the arctic cod, the North Sea herring, or the yellowfin tuna. The inquiry needed is one based on the disciplines of fisheries biology.

The responses of fish stocks to climatic factors and to climatic changes reveal much of their essential biology. They were linked by Ottestad to the same factors that govern the widths of the annual rings on pine trees. The only way in which such obvious atmospheric factors can be linked to the fish stocks in a simple way is through those factors that affect the production cycle — wind strength, wind direction, and solar radiation.

The responses of fish stocks to long-term changes with a period of a century or so can be accounted for in a specialized form of the stock and recruitment relationship. The responses to annual changes, as differences in recruitment from year to year, may be well accounted for in a speculative manner by the match, or mismatch, or larval production to that of their food. Such a mechanism may be an expression of the single process put forward in the last chapter, by which recruitment and stabilization are achieved at the same time. A description of the Russell cycle reveals that analogous effects may be distributed throughout the marine ecosystem; indeed, there is a suggestion of an alternative cast of actors waiting off stage during any one climatic period.

10

The Future of Fisheries Research

The first nine chapters represent a brief review of fisheries science to the late 1970s. This chapter is called the "future of fisheries research," although the future is hard to forecast. It is a view of what might happen — a personal view, much influenced by my friends on the Suffolk shore and in many parts of the world.

Migration

Harden Jones (1968) not only summarized our knowledge of fish migration in the sea until then, but also revolutionized the study by noting that fish in the midwater can rarely have an external referent. Therefore they may be unable to navigate and must drift with the currents. Until the last decade or so oceanographers have described the major ocean currents and their variability. Recently, however, as current meters, Swallow floats, STDs (salinity, temperature, and depth recorders), and expendable bathythermographs have been deployed in greater numbers at more accurately determined positions, they have shown that the major currents have spawned large eddies and permanent countercurrents. For example, Kvinge, Lee, and Saetre, (1968) have described elongated swirls in the relatively warm Atlantic stream off the Norwegian coast. The countercurrent below, and inshore of, the Gulf Stream is well known (Knauss, 1969), but currents and countercurrents are now being discovered off the west coast of the British Isles.

Plaice and cod use selective tidal transport in the southern North Sea, a shallow region of turbid water with very strong tides (Harden Jones et al., 1979). This form of migration was discovered with the use of the Admiralty Research Laboratory sector scanner. Figure 101 summarizes the evidence for selective tidal transport with the use of the scanner. In the future, the modes of migration will be investigated in the somewhat deeper waters over the continental shelf, where the tides are weaker and the residual currents relatively more pronounced. Tagging experiments have long been used to describe the spread of fish from the point of liberation, and models have been developed which account for such spreading with or without directional biases. Selective tidal transport and its analogues in deep water may indicate the behavioral mechanisms by which such directional biases can be quantified, and tagging experiments may be designed to test them. If the spread of a stock throughout the year on its annual migrations were described statistically, it would represent the seasonal course of the fisheries themselves.

Fish have considerable capacity for olfaction. That capacity is probably used to the full by the sockeye salmon as they swim up the Fraser River in British Columbia, back to the parent stream, by sequential homing (Hasler, 1966; Cooper et al., 1975). It is a significant discovery not only in the description of sockeye migration, but because other fish species may make use of similar olfactory clues to reach their restricted spawning grounds. The Arcto-Norwegian cod may ride a countercurrent from Svalbard to the Lofoten Islands, as Harden Jones suggested, but we do not know how they reach the narrow strip on the edge of the deep water where they spawn, not far from the port of Svolvaer.

The Downs herring may migrate into the Southern Bight of the North Sea by a form of selective tidal transport, but we do not yet know how they reach their fixed and restricted spawning grounds like that near the Sandettié Bank.

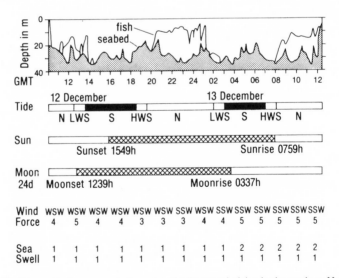

Figure 101. The vertical movement of an acoustically tagged plaice in the southern North Sea with respect to the tidal streams. The fish remains on the seabed on the southbound tide and swims in midwater on the northbound one. There is no correlation with sunset or sunrise, moonset or moonrise, or with weather conditions. Wind force and sea swell are on the Beaufort scale. Adapted from Harden Jones et al., 1979.

Cod may be led by runoff from the glaciers to the "fish-carrying layer" at 70 m in the Vest Fjord. Herring may be led to the gravels on the edges of some banks in the southern North Sea. Harden Jones believes that they might be homing on to sources of groundwater seepage. Both runoff and seepage may carry molecules which remain the same from year to year.

The importance of these discoveries to the future direction of fisheries research lies first in the biological description of a unit stock and second in the description of fisheries as mobile distributions of fishing effort. The early fisheries biologists and many of their successors reached common-sense solutions to such problems. But the extension of the use of quotas to small stocks in small areas requires that the scientific basis of stock analysis be understood independently of the fishery.

Unit Stock

Twenty years ago it was still possible to describe a unit stock by the homogeneity of the vital parameters of recruitment, growth, and mortality, which almost implies that different species could be grouped together in one "stock" provided that those vital parameters were the same. At that time recruitment was considered independent of stock size at observed levels of exploitation. Because so many pelagic stocks have suffered from recruitment overfishing since the early 1950s, stock unity must be established independently. Genetic methods have shown that stocks of North Atlantic cod are distinct from their neighbors. In view of the clines and continua in those populations studied on the land, this remains a somewhat surprising result.

By analogy with fish tags, the blood proteins such as transferrins have sometimes been called genetic markers, which is a misleading description because the distributions of genes and tags yield different forms of information. Samples of transferrins taken on a spawning stock describe the genetic structure of that stock as distinct from that of a neighbor. Tags released on the spawning ground are recovered across the area occupied by the stock — provided, of course, that fishermen explore that area. The distribution of recoveries describes the area occupied by the stock during its annual migration. If enough are released, some tags will be recovered outside the normal range from strays or colonizers which will never return. The two methods may be combined by taking blood samples of tagged fish released on spawning grounds. Then the genetic structure can be described as a distribution in area rather than a mere difference between well-chosen samples. The degree of mixture on common feeding grounds could be estimated from the proportion of tags of each spawning group in the area of common origin. The genetic structure in the area of mixing may then be contrasted with that in those areas where mixing does not occur.

Li (1955) noted that in any large population a number of spawning groups are needed between which exchange is low in order to maximize the variance of the whole. Each spawning ground may be analogous to the parent stream of the Pacific salmon, and the stray from stream to stream or from ground to ground may be the exchange needed to obtain the high variance. There are three or four spawning grounds of cod in the North Sea (Daan, 1978), but no genetic differences have yet been established between them. De Veen's (1961) tagging experiments on plaice in the southern North Sea showed that the spawning fish returned to the same ground each year with very little stray. Yet Purdom, Thompson, and Dando (1976) have detected no genetic differences between the three spawning groups. The tagging experiments and the genetic observations yield different forms of information.

The history of racial analysis in fisheries biology goes back to the studies of Heincke (1898) on the morphometric measurements on herring from different areas in the North Sea. They were not successful, but the object of this work was to isolate the populations on which subsequent analysis might have been based. Then the only form of overexploitation understood was recruitment overfishing; indeed Petersen (1894) thought that growth overfishing would be solved by fishermen themselves in response to economic pressures. The need for highly developed racial studies reappeared when recruitment overfishing was "rediscovered," because in stock-recruitment studies we need to describe that stock which generates the recruiting year classes. It is reasonable to

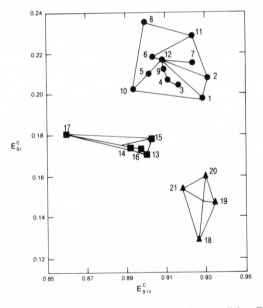

Figure 102. The relationship between the gene frequencies in two alleles, E^C_{SI} and E^C_{SIV}, in samples from three groups of North Sea herring: Downs, Dogger, and "coastal." The numbers indicate the station numbers on the cruises. Adapted from Zenkman and Lysenko, 1978.

suppose that the cod stock in the North Sea generates recruiting year classes from the three or four spawning grounds that are homogeneous. Theoretically, if not practically, the catches of fjord cod in the Vest Fjord should be separated from those of the Arcto-Norwegian cod.

In the preliminary studies conducted on North Sea herring races in the 1950s, the Downs and Dogger stocks were separated partly with meristic characters but more convincingly with uncorrelated year classes. The stocks collapsed in sequence during the 1960s, as did the remaining groups in the mid-1970s; catches of herring are now banned from the North Sea. Zenkman and Lysenko (1978) have separated samples of Downs and Dogger fish with the gene frequency of two esterase alleles (Fig. 102). The samples were separated independently by egg sizes and meristic characters. If this work is confirmed and the stock recovers in parts, then the subsequent management should be restricted to those parts.

Growth and Mortality

The growth of larval and juvenile fishes has rarely been well described because until recently they have not been readily aged. Further, there appears to be a break in growth rate at metamorphosis; for example, cod larvae grow at 10 percent per day and the little post-metamorphosis fish grow at 2.5 percent per day (Cushing and Horwood, 1977). In other words, growth before metamorphosis differs profoundly from that after it. The von Bertalanffy equation well describes post-metamorphic growth on a scale of years. Another form might describe pre-metamorphosis growth in days, much as Zweifel and Lasker (1976) have done. If daily rings on larval otoliths can be read easily in fishes other than the anchovy off California, considerable advances in our understanding of larval growth will ensue. If we are to understand how recruitment is generated from parent stock, we must understand how biomass is created by the growth of individuals in face of considerable loss in numbers.

Another facet of the same problem is the study of density-dependent growth. Preliminary studies suggest that the growth of adult marine fishes is not density dependent, but that the growth of juvenile or larval fishes does depend upon their density. So far, such little fishes have been poorly sampled in the catches of the adults, for the bigger members of an age agoup may be caught but not the smaller ones. If the growth of the younger fish is density dependent and that of the adults is not so, then there must be an age at which such density dependence is extinguished. If growth and mortality are linked, that age is of considerable importance in studies of stock and recruitment.

In larval and juvenile life, growth may depend on density until numbers are reduced to that level at which the animals may grow unimpeded. They die by predation in the main, although as proximate cause predation may represent other sources of mortality. In later and unexploited life some fishes may die by predation, but others must die of disease or old age. One problem facing fisheries biologists is the estimation of natural mortality independently of fishing mortality. The component due to predation can, in principle, be estimated quite properly; Figure 94 (Chapter 9) shows the quantities of pilchards and maasbanker taken by predators off southwest Africa. The ratio of catches to the sum of catches and quantities eaten estimates the exploitation rate, F/Z. A study of this problem would require full knowledge of the predatory net that exploits a given fish population, which demands considerable sampling effort, such as that initiated by Grosslein, Langton, and Sissenwine (in press).

If such sampling effort were extended to all stages of the life cycle, the relationship between growth and mortality might become elucidated. In order to create biomass the growth rate must be greater than the death rate, yet both decline with age, and the rate of decline diminishes with age until the critical age is reached when growth rate and mortality rate become equal. It is possible that the ratio $(G'/M) > 1$ remains more or less constant from the time of first efficient feeding until the critical age. Let us recall that a predator weighs about a hundred times more than its prey and then extend the rule throughout the life history of a fish. Figure 35 (Chapter 4) shows the logarithmic growth curve of weight on age for plaice. We might imagine two other curves, that of mythical predators scaled two orders of magnitude above the weight curve of plaice, and that of mythical prey scaled two orders of magnitude below it. The three curves would represent the life-long process of fishes through all the trophic levels in the sea. For any interval of time, the growth rates of all three trophic levels are the same. Any growth increment in one trophic level is a linear function of the decrement in numbers in the trophic level below, from which it follows that growth and mortality are linked in the middle trophic level, that is, in the life history of the fish.

More generally, fisheries biologists may extend their studies beyond the single populations which have been so well documented for so long. Indeed a more extended study is sometimes needed to understand the changes in a single population. One of the effects of successive El Niños during the period of the anchoveta fishery off Peru was to reduce the number of predatory guanay birds from 16 million in 1957 to 4 million after the 1965 event, and then to one million after that of 1972–73. The recruitment to the anchoveta stock increased from the period 1960–64 to that of 1965–70 by a factor of 1.6. The decline in predation and the increase in recruitment may well be associated. Then natural mortality of the recruiting anchoveta might have declined and stock might well have increased until the collapse in 1972. Perhaps this increment was undetected in the study of the single population and was only revealed when the biological effects of El Niño were revealed.

Much ecological research is concerned with ecosystems or communities rather than with single populations. There is a false polarization between the energy transfers or numerical taxonomy of ecosystem research and the studies of single stocks in fisheries research. Between the two lies a somewhat unexplored region which includes the links between a few specified populations, particularly in functional terms. For no single population is the vertical trophic structure of significant competitors at any stage of the life history of any single population known in any complete way. Such information is needed before the position of a fish population within the trophic structure can be understood.

The Measurement of Abundance

Stock density, or catch per unit of effort, is theoretically a good index of stock — e.g., $F = C/N = qf$; $C/qf = N$ (the same conclusion is given in Chapter 5, in estimates averaged throughout the year). However, the well-known inverse relationship of stock density upon fishing effort (introduced by W. F. Thompson) may be biased in two ways: (1) the increase in power or efficiency as fishermen learn to catch more effectively; (2) the variation in catchability with season and ground. The capacity of fishermen to learn represents the most destructive bias in the inverse relationship between catchability coefficient and numbers. The bias can be detected with the use of concentration factors, as described in an earlier chapter, but its elimination would be effective if independent methods of stock estimation were available.

Additional problems arise in mixed fisheries because any trawl haul comprises a number of species and their stock quantities will all differ. The effort exerted in that haul is, of course, common to all species, but each catchability coefficient should estimate the appropriate stock quantity. Then the effort is randomly distributed, which might well be true for a single haul. A skipper can only guess the catch on the first haul of a trip, but by the last he knows what he will catch. Hence, the effort from haul to haul is not randomly distributed, and the catchability coefficient of the preferred species will be biased upward. There are two ways in which the problem of preferred species in a mixed fishery may be approached. First, a record of catches of each species in a single haul, for example, a skipper's log, might indicate the species at which the trawl was directed. The catchability coefficients might then be partitioned between preferred species and the others; data on the former would then be processed with concentration factors and on the latter without them, because the effort is distributed randomly on the species caught inadvertently. A second method of approaching the problem of directed catches is provided by acoustic methods, which may in the future estimate stock density as numbers, or biomass/km^2, independently of the catching gears.

Acoustic methods are used for this purpose (Cushing, 1973b), but the integrators employed at the present time are open to a potential bias; the lowest signal or threshold may be received from either a given number of the desired targets or from a larger number of smaller targets of no interest

(Cushing, 1978a). For example, the stock of blue whiting (*Micromesistius poutassou Risso*) to the west of the British Isles lies in a layer about 60 miles wide in 400–500 m off the continental shelf. When a midwater trawl is towed through the layer, only blue whiting are caught, but pearlsides (*Maurolicus mulleri* Gmelin) are often taken in the trawl cover; they represent a small overestimate to the estimate of stock. As the trawl is hauled frequently merely to identify the target, the bias here was identified and eliminated.

For a fisheries biologist the ideal acoustic method would be a fish counter which also sized the targets because the threshold signal can represent the smallest fish required in the sample. Then the potential bias in the integrator is removed. It is unlikely, however, that a fish counter can be designed which will resolve all shoals into single targets. We therefore need to be able to discriminate between single fishes and shoals.

B. J. Robinson (1978) has devised a method of measuring target strength at sea. The transducer was lowered to within 10 or 20 m of the layer of blue whiting, and a large number of observations were recorded on tape. Single targets were isolated with somewhat severe criteria; they could be separated and counted, but would represent an underestimate of density merely because the criteria were severe. However, the same system can integrate the signals within a range gate and from transmission to transmission. Then, within a sampling unit of many transmissions, we would have signals from single fishes and an integral of all signals received. The single fishes yield an estimate of target strength, which can be used to convert the integral to numbers of fish. This technique can be developed to take a large number of target strength estimates on each survey.

Any system that discriminates between single fishes and shoals requires an assumption of identity. That is, within the unit of sampling the single fishes and those in the shoals are fish of the same species and of the same size; then the bias in the integrator is not eliminated but is considerably reduced. The single fishes can be formed into a size distribution between sampling units as well as estimates of target strength within them. Such a system would provide the basis for an independent method of estimating stock and stock density at sea.

Acoustic methods have been used successfully for exploratory surveys. The biases in catches per unit of effort are perhaps best corrected by independent methods of estimation, because they probably cannot be eliminated by the mere manipulation of statistics. The success of cohort analysis has led some to believe that estimates of stock density are no longer needed, but a step is required from the historical description of the stock, which is cohort analysis, to its management in the future, which is the quota. Already, independent stock estimates are used for this purpose, but as a science advances it needs an interlocking system of independent estimates.

Stock and Recruitment

The dependence of recruitment on parent stock is the central theoretical problem of fisheries biology. The formulations of Ricker (1954, 1958, 1973) and of Beverton and Holt (1957) account for two current hypotheses and provide convenient statistical summaries, but because recruitment is variable the relationship has often been described only after it is too late — when recruitment overfishing has supervened. Both the published formulations depend upon a separation of the mortality between hatching and recruitment into density-dependent and density-independent forms; from observations an average statement of the relationship can be derived. But there is no clear theoretical basis such as is provided by the yield-per-recruit formulation in the simpler solution to the problem of growth overfishing.

Two steps have been taken toward the establishment of the theory needed. The first was the suggestion in Chapter 7 that the dome of a stock-recruitment curve might be generated by cannibalism even if the basic mechanism depended upon the growth capacity of larvae and of juvenile fish (Cushing and Horwood, 1977). Hence the dome is perhaps a special case limited in the sea to the gadoids. A second step has been taken by John Shepherd (Shepherd and Cushing, 1980): if fish stocks can sustain exploitation by a factor of 3 or more in (F/M) for long periods of time, the density dependence must be strong. As a factor of 3 or more cannot result from what we imagine of density-dependent fecundity, it must be effected by density-dependent mortality in larval or early juvenile life when mortality is high. Yet the perch in Windermere control their population by density-dependent fecundity (Le Cren, 1962); perhaps stocks in the sea make use of fecundity as a trimmer to the main system.

Ricker (1958a) and Ricker and Foerster (1948) suggested two biological mechanisms. The first is the aggregation of predators, which could generate mortality dependent upon stock differences from year to year. Cushing (1955) showed that herring aggregated on to patches of *Calanus* in about three weeks, after which they disengaged. Their prey was more abundant than fish larvae by two orders of magnitude and hence aggregation by the same number of predators would be very much slower, too slow to generate the differences observed. The second mechanism proposed is the increase in mortality with abundance in a predatory field when food is potentially limiting. Harding, Nichols, and Tungate, (1978) observed that the differences between the mortality rates of plaice eggs and the subsequent larvae in the same year class are much less than the overall rates from year to year. The mortality rate may vary independently of larval mortality, but the cumulative mortality is perhaps density dependent. Let us suppose that food were short for a period such that growth of the larvae were density dependent; then the cumulative mortality during that period would also be density dependent. One might imagine that

as the animals encounter patches of their food they would grow at maximal
rate, but more slowly between patches, as the quantity of food per larva was
reduced there. Then growth and cumulative mortality would be density dependent when food is short, density independent when the larvae grew at
maximal rate. Thus the density-dependent and density-independent processes
would only be separated in some form of stochastic analysis.

Shepherd and Cushing (in press) developed an explicit formulation of
density-dependent growth which resembles Ivlev's earlier model (1961):
$G'(N) = G'_{max}/[1 + N_0/K'']$, where G_{max} is the maximal growth rate; N_0

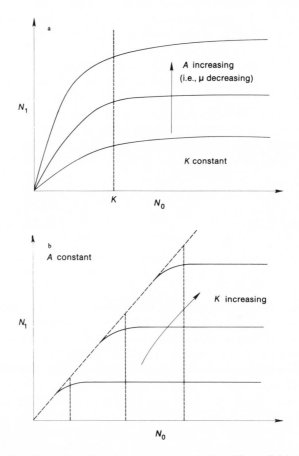

Figure 103. Dependence of recruitment upon parent stock developed from a link between growth
and mortality as described in the text. N_1 is recruitment in numbers and N_0 is the number of eggs;
A is the fractional survival due to density-independent causes, and K is an index of food availability. In (*a*) is shown a family of curves with increasing A and constant K; in (*b*) is shown a
family with constant A and increasing K. Adapted from Shepherd and Cushing, in press.

the initial numbers, and K'' an index of food abundance. During a critical period, larvae grow through the predatory field and growth is density dependent; the cumulative mortality includes both components, density dependent and density independent. But with increasing larval numbers, growth depends more on density, and the density-dependent component of the cumulative mortality becomes greater. The equation developed to model these ideas in a nonstochastic manner is:

$$N_1 = A_2 N_0 / \{1 + [(1 - A_2) N_0 / K'']\}, \tag{152}$$

where N_0 is the number at the start of a critical period,

   N_1 is the number at the end of a critical period, and

   A_2 is the fractional survival due to density-independent causes,

   $= \exp [- (\mu / G'_{max}) \ln (W_1 / W_0)]$,

where W_0 and W_1 are initial and final weights;

   μ is the instantaneous mortality rate.

A stock-recruitment curve is formed as illustrated in Figure 103, with A_2 the slope at the origin and an asymptote $A_2 K'' / (1 - A_2)$. Although, by and large, A_2 may be characteristic of groups of fishes, such as clupeids, flatfishes, or gadoids, the mortality rate, μ, must be expected to vary from year to year if the results of Harding, Nichols, and Tungate (1978) on the plaice have general significance. The curve is a function of food, predatory mortality, and initial numbers, and the three parameters make it unique for each year class. However, an averaged curve can be formed with many observations and would be used in management as frequently as any other stock and recruitment equation. Much depends upon the ratio (G'/G'_{max}) and how variable it is; from Ware's work (Fig. 49, Chapter 4) it is possible that differences in μ from year to year are less than the differences in the ratio between groups of fishes. If it were true, then Equation (152) might be used for management purposes without observations in much the same way as is the yield-per-recruit formulation.

Equally important, or more so, are the potential tests of the hypothesis. The simplest one might be the time to reach metamorphosis as index of the critical period; this time certainly depends on temperature, may be influenced by endocrine events, and perhaps is determined by the availability of food. A more complicated test depends on the precise aging of larvae either by sampling the same population in a patch at sea or by reading the daily rings on the larval otoliths. Walsh (1975) developed methods to describe an upwelling plume well enough to predict its biological and hydrodynamic course with programmed hypotheses which can be subsequently tested by observations, all on a single voyage. Such methods could be used to test the expectation that the growth of fish larvae is density dependent as a function of available food.

The scientific problem of stock and recruitment has advanced to the point at which hypotheses might be tested. But the problem of managing recruitment overfishing remains; if estimates of recruitment and parent stock are available, judicious use of the resulting scatter diagram can prevent overfishing from taking place. The essential step is to realize that the scatter does not establish the independence of recruitment from parent stock and that it may mask the true relationship.

Management

Two models are used in management: the descriptive one due to Graham (1935) and Schaefer (1954), and the analytic one due to Beverton and Holt (1957). The descriptive model is based on Thompson's rule, which relates the decline of stock density to increase in fishing effort, and an estimate of maximum sustainable yield can be derived from it. However, if catchability increases inversely with abundance, undetected, the slope of Thompson's rule is less than it should be and the maximum sustainable yield is overestimated. With the analytic model, stocks are reconstituted from estimates of their vital parameters. Since 1970, quotas have been introduced in the North Atlantic; both models were used to establish them. But the widespread use of cohort analysis, which might have been designed to calculate quotas, has demanded the analytic model; indeed, yield-per-recruit solutions are readily reached with cohort analysis.

Any cohort analysis at the end of the last year is used in the present year to estimate a quota next year. As shown in Chapter 5, however, the analysis is unreliable for the three previous years, so there is a gap of five years between the entirely reliable data and the quota. Of course from year to year, as the analysis is rerun with new material, much of the possible unreliability is reduced, but there is often a good case for estimating stock density in the current year independently; for this purpose changes in fishing effort from year to year, estimates of stock by tagging experiments, and egg surveys have been used. Perhaps in the future groundfish surveys or acoustic surveys will be used for the same purpose. Stocks vary from year to year due to differences in recruitment, and so the annual quotas change from year to year. In the North Sea, a young fish survey is carried out each year, in February, by the International Council for the Exploration of the Sea, in order to estimate the annual recruitment to certain stocks. Groundfish surveys may become part of the international management structure in the future.

A stock exploited at maximum or optimum sustainable yield will vary naturally from year to year, and annual quotas would vary in the same way. The industry which exploits that stock needs long periods in which to replace its vessels or train its skippers. Hence, in addition to an annual quota which will be needed to maintain enforcement, a long-term quota would be desirable. For example, one might estimate the maximum sustainable yield of the plaice in the southern North Sea, but from our past experience an outstanding year class like that of 1963 might raise the stock by 20 percent or more for a number of years. The variability of quotas due to ordinary year classes might be less than 10 percent. Further, it should become possible to forecast the annual quotas as the results of young fish surveys become more generally applied. But such arrangements will only become feasible when stocks are exploited at the level of the desired management objective.

It has long been known that the effects of fishing could be controlled by restraining fishing effort. The great advantages are that the number of ships is controlled explicitly for a number of years and the annual variations due to recruitment can be ignored. Michael Graham put forward this method of control at the Overfishing Convention of 1946. Cohort analysis can be manipulated to convert catch at age in numbers to stock at each age and year, or to fishing mortality at each age and year. Then as $F = qf$, it should be possible to estimate the effort needed to attain a desired management objective. However, there are two present bars to such a development; the first is the pitiful lack of adequate statistics on effort in the Northeast Atlantic, and the second is the inverse dependence of the catchability coefficient upon abundance. In time, both difficulties will be overcome.

In some areas licences are issued by the coastal state to vessels, and the number of licences is established more or less arbitrarily, determined by the contemporaneous catch rates of those vessels. From this arbitrary base line a form of effort control might well develop as proportionate decrements or increments are introduced in accordance with whatever management objective is needed. This rough form of effort control might indicate the entry of more developed methods.

One of the more intractable problems facing a manager of fisheries is that of the mixed fisheries. Very few stocks are exploited in isolation, if only because no skipper can be sure what he will catch. In the Barents Sea, cod and haddock comprise the major catches but the stock quantities differ, so haddock remains heavily exploited while the cod is brought slowly toward the maximum sustainable yield.* Similar problems arise with plaice and sole and separately with cod, haddock, and whiting in the North Sea. But some fisheries are much more mixed. An extreme case is that in the Gulf of Thailand, where up to sixty species are exploited to more or less the same degree. Brown et al. (1975) has used a Schaefer model to estimate the maximum sustainable yield for "total finfish" in the Northwest Atlantic (excluding the catches off West Greenland), as Brander (1977) has done for the Irish Sea. Both presentations are much clearer than the usual Schaefer plots for single stocks. But the catchability coefficient really refers to the total catches of the trawl, and the proper partitioning of catchability coefficients by stocks in such mixed fisheries has not yet been achieved.

Some believe that fisheries science should be limited to the accountancy needed only for management purposes. No more is needed to use a method, but if such an attitude presupposes confidence, it is limited. A science builds itself with interlocking steps of evidence, each independently derived. We are permanently in need of new steps and methods merely to make the science.

*Note: this is no longer true, because it has been recently shown that the Arcto-Norwegian cod stock is too heavily exploited.

Reference Matter

List of Symbols and Their Definitions

A Area (m^2, km^2, or statistical rectangles).

A' Coefficient in Beverton and Holt's stock and recruitment equation (Equation [128]).

A_r Maximum cross-sectional area in Gray's equation describing the resistance to a rigid model of a fish (Equation [115]).

A_1 Maintenance threshold in Birkett's equation describing the dependence of growth increment upon food (Equation [24]).

A_2 Fractional survival due to density-independent causes.

B Number of births per year.

B' Coefficient in Beverton and Holt's stock and recruitment equation (Equation [128]).

C Catch in numbers.

         C_i = catch in numbers in the ith year.

D True density of stock.

         D_0, D_1 = true density in year 0, year 1.

D' Number of deaths per year.

D'' Genetic distance.

E Rate of exploitation.

         E_i = exploitation rate in ith year.

         E_n = exploitation rate in nth year.

E' Echo level in dB.

E'' Number of emigrants per year.

F Instantaneous coefficient of fishing mortality.

         F_i = instantaneous coefficient of fishing mortality in the ith year.

         F_n = instantaneous coefficient of fishing mortality in the nth year.

         F_s = instantaneous coefficient of fishing mortality at which the maximum sustainable yield is generated.

         $\hat{F}$ = estimate of instantaneous coefficient of fishing mortality.

G Growth increment.

G' Specific growth rate.

         G'_{max} = maximum specific growth rate.

H, H'' Fraction of tagged fish that survives tagging.

H' Food needed to satiate in g; increment of food.

H'_0 Fullness of gut needed to satiate in g.

H'_k Maximum capacity of gut in g.

H'_{t_0} Quantity of food needed to satiate when search starts, or attack threshold, in g.

H'_l Quantity of food needed to satiate when prey is captured in g.

H'_{t_l} Attack threshold when prey is captured.

H^* Increment of food.

I Echo level in acoustic intensity in watts per m^2.

I_0 Source level or transmitted acoustic intensity in watts per m^2.

I'_0 Intercepts in equations describing returns of tags in time.

I'' Number of immigrants per year.

J, J' Fraction of total recaptures reported.

K Rate at which length reaches L_∞, the asymptotic length in the von Bertalanffy
                equation (Equation [14]).

K'_1 Efficiency of feeding.

K'_2 Efficiency of absorption.

K'' An index of food abundance.

K^* Carrying capacity in the logistic equation (or food available to the population)
                (Equation [141]).

L Length in cm.

        L_m = length at maturation.

        L_∞ = asymptomatic length in the von Bertalanffy growth equation (Equation [14]).

M Instantaneous coefficient of natural mortality.

        M_1 = instantaneous coefficient of density-independent natural mortality.

        M_2 = instantaneous coefficient of density-dependent natural mortality.

        M_0 = natural mortality at the start of a cohort's life, i.e., at hatching.

N Number of fish.

        $N_0, N_1, N_2, N_t, N_\lambda$ = stock at the end of year 0, year 1, year 2, year t, and year λ.

        N_c = number of fish larvae at t_c, the critical time.

        N_i = number in stock at the beginning of the ith year; number tagged before
                ith period and alive at the end of it, in a multiple tagging experiment.

        N_i' = number of tagged fish in the sea at the beginning of ith period.

        N_m = number of fish tagged.

        N_{m_i} = number tagged in the ith year or the ith week; number released at the end
                of period i, in a multiple tagging experiment.

        N_m' = number of fish tagged on second occasion.

        N_0 = number of fish at t_0, time of hatching; prey density, in Holling's analysis
                of predation.

        N_r = number of fish recaptured.

        N_{r_i} = number of fish recaptured in the ith time period.

        N_t = number of fish at time t.

        N_{t_i} = number of fish tagged.

        $\bar{N}$ = average abundance or stock density.

        $\bar{N}_1, \bar{N}_2$ = average stock density in year 1, year 2.

        $\bar{N}_i$ = mean number of fish tagged in area i.

O Average number of eggs per m^2.

P Stock in weight.

        P_1, P_2 = stock in weight in year 1, year 2.

        P_m = maximum stock in weight; maximal stock in Ricker's stock and recruitment
                equation (Equation [131]).

        P_w = average stock in weight.

P_e Number of eggs spawned.

P_n Stock in numbers.

        P_{n_i} = stock in numbers in the ith period.

P_r Replacement magnitude of stock.

P_s Stock from which the maximum sustainable yield is taken.

Q Temperature coefficient of egg development.

Q' Energy.

         Q'_a = energy of swimming activity.

         Q'_c = energy in the daily ration, r.

         Q'_d = energy of specific dynamic activity.

         Q'_g = energy in the increment of weight.

         Q'_s = energy in standard metabolism.

         Q'_w = energy in daily excretion.

Q'' Constant relating natural mortality and growth in Ware's equation (Equation [37]).

R Recruitment; increment of recruitment in one year.

R' Number of recruits entering a fishery.

R'' Increment in weight due to recruitment.

R_3 Stock density of fish as at three years of age.

R_m Maximal recruitment.

R_r Number of recruits from the replacement stock, P_r.

R_s Resistance of a rigid model in kg.

S Survival.

         S_i = survival in the ith period.

         S_n = survival from year to year in cohort (or virtual population) analysis.

         S_r = survival rate at P_r.

S_s Strike success.

S' Source level in dB.

S'' Constant in an equation that separates type A and type B errors in a tagging experiment (Equation [103]).

S^* Percentage of maximal mortality in calculation of total allowable catch from cohort analysis.

T Temperature in degrees Celsius.

T' Metabolic rate in ml O_2 per g body weight.

T'' Target strength in dB.

T^* $N_1 + 2 N_2 + 3 N_3$ in Chapman and Robson's survival equation (Equation [86]).

U_i Percentage of ingredient i in a food selection experiment.

U_j Percentage of ingredient j in a food selection experiment.

V Velocity; swimming speed in cm per sec or lengths per sec.

V_n Virtual population at age n in cohort analysis.

V_r Difference between speeds of predator and prey in cm per sec.

V' Velocity of fish in Gray's equation in m per sec (Equation [115]).

V''_p Predator's attack speed in cm per sec.

W Weight.

         W_∞ = asymptotic weight in the von Bertalanffy growth equation (Equation [16]).

         W_c = weight at first capture.

         W_0, W_1, W_t = weight at time 0, 1, t.

W' P/P_m in Ricker's stock and recruitment equation (Equation [131]).

X Other loss coefficient.

Y Yield in weight.

Y_{ij} Catch in the ith rectangle in the jth period.

Y_s Maximum sustainable yield in weight.

Z Instantaneous coefficient of total mortality.

         Z_1, Z_2 = total mortality in first year, second year.

         Z_c = instantaneous coefficient of density-dependent mortality.

         Z_i = instantaneous coefficient of density-independent mortality.

Z_r = mortality from spawning to recruitment.

Z_x, Z_y = instantaneous total mortality coefficients in two fisheries.

Z' Decrement in weight due to total mortality in Russell's equation (Equation [1]).

a P_r/P_m in Ricker's stock and recruitment equation (Equation [131]); homozygote, in the Hardy-Weinberg law.

a' Rate of natural increase in Schaefer's logistic equation (Equation [50]) and its derivatives.

a'' Constant in the Paloheimo and Dickie growth equation (Equation [20]).

a^* Proportion of food absorbed in the coefficient of food absorption.

a_1 Dispersion coefficient.

a_1'' Constant in an equation relating the total mortalities in two fisheries (Equation [5]).

b Relative birth rate per year; homozygote, in the Hardy-Weinberg law.

b_1 Index of density dependence; power function in Ware's equation (Equation [37]).

b' Constant in the Paloheimo and Dickie growth equation (Equation [20]).

b'' Constant in the equation that describes the trend of natural mortality with age (Equation [120]).

b^* Directivity coefficient in the description of a sound beam (Equation [82]).

c Constant of integration.

c' Speed of sound in sea water in m per sec.

c'' l_c/L_∞.

d Estimated stock density or catch per unit of effort in numbers.

d_0, d_1 = stock density in successive years.

d' Relative death rate per year.

d' Scaling constant in Richards's growth equation (Equation [17]).

e Relative emigration per year.

f Fishing intensity, or fishing effort per unit area.

f_i = fishing intensity in area i.

$\bar{f}$ = effective overall fishing intensity or concentration factor.

f' Specific rate at which the gut is emptied.

f^* Number of eggs per adult.

g Fishing effort, or time spent fishing.

g' Gravity in cm per sec per sec.

h Ratio of schools caught cooperatively and those taken independently.

i Relative rate of immigration per year.

k Number of stations in a statistical rectangle.

k_1 Coefficient of predation.

k_3 A dimensionless coefficient in Gray's equation describing the resistance of a rigid model of a fish (Equation [115]).

k' Scaling factor in the Parker-Larkin growth equation (Equation [18]).

k'' Feeding coefficient in the Ivlev equation (Equation [28]).

k^* Constant in specific growth rate equation (Equation [34]).

k_1^* Scaling constant.

k_2^* Constant in equation describing the trend of natural mortality with age (Equation [116]).

l Length in cm.

l_c = length at first capture.

l_t = length at time t.

l_1 = length at the end of the first year's growth in herring.

m Proportion of larvae in the surface layer.

m' Power in Richards's growth equation (Equation [17]).

m''_0 Initial mortality rate in Gompertz's equation (Equation [121]).

m''_1 Rate of increase of mortality with age in Gompertz's equation (Equation [121]).

$m*$ Power in Pella and Tomlinson's equation (Equation [148]).

n Number; number of fish recovered; numbers 0–3 in summation constant.

     n_1, n_2 = numbers of fish recaptured in year 1 and year 2 in a tagging experiment.

     n_3, n_4, n_5 = numbers of fish per unit of effort recruiting at ages three, four, and five.

     n_i = number of fish recaptured.

     n'_i = number recaptured in a length group different from that in n_i.

p Probability of capture.

p' Concentration of food.

$\bar{p}$ Mean food density.

q, q' Catchability coefficient.

     q_0, q_1 = catchability coefficient in year 0, year 1.

$q*_i$ Recaptures during rest of their lives of fish tagged before the ith period.

r Daily ration in weight.

     r_{max} = maximum daily ration.

r_i Recoveries of tags during the rest of their lives from the number tagged.

r_n Ratio of stock at the beginning of one year to the catch in the previous one, in cohort analysis.

r^2 Coefficient of determination.

r'_n Shortest distance traveled by tagged fish.

s Mean angular deviation in radians.

s^2 Variance.

t Time; days; years; days at liberty in a tagging experiment.

     t_c = age at first capture; time to reach the critical size at which a larva leaves a predatory field.

     t'_c = time required to capture and eat a prey.

     t_0 = time of hatching; age at which growth appears to start, for statistical reasons, in the von Bertalanffy growth equation (Equation [14]).

     t_p = time spent in pursuit.

     t_r = age of recruitment to the spawning stock.

     t_s = searching time.

     $t\alpha$ = digestive pause.

     $t\lambda$ = age at which the year class is extinguished.

u percentage of an ingredient, i, of food in the gut.

v Percentage of an ingredient, j, of food in the gut.

v' Directional component of velocity.

x_i Gene frequency.

y_i Gene frequency.

Greek Alphabet

α Density-independent survival.

α' Searching coefficient.

α'' Scaling constant in equation relating catchability to numbers.

$\alpha*$ Scaling constant in metabolic equation; mean angle to the coastline at which tags are recovered.

β Coefficient of density-dependent mortality.

β' Power relating catchability to numbers.

β'' Constant determined by the shape of the search field in Holling's equation (Equation [30]).

γ Daily ration at zero aggregation.

δ Power in metabolic rate equation.

δ' Constant in Ware's equation relating growth rate to mortality rate (Equation [37]).

δ'' Constant in Gulland's yield equation (Equation [149]).

$\delta*$ Scaling constant in Ware's growth-rate equation (Equation [37]).

ϵ Constant in Gulland's yield equation (Equation [149]).

ζ Index of aggregation.

η Index of food selection.

θ Bearing of a recaptured tagged fish measured in angle from the axis of the coastline.

θ' Angle from the acoustic axis of a sound beam.

θ'' Daily ration as percentage of maximum ration at zero aggregation.

κ Feeding coefficient.

λ Fishable life span in years between entry to the fishery and extinction.

μ Coefficient of larval or juvenile mortality.

μ_1 Coefficient of density-independent mortality.

μ_2 Coefficient of density-dependent mortality.

ν Deviation in density from the mean of a series in space.

ξ Efficiency of food conversion in units of nitrogen.

ρ Density of sea water.

ρ' Exchange ratio per tidal cycle.

σ Scattering cross-section in m^2.

τ Equivalent time interval (used in results from tagging experiments).

τ' Duration of acoustic pulse.

ϕ Translocation rate.

ϕ' Azimuthal angle of a sound beam.

χ Concentration coefficient.

ψ Number of tidal cycles per day.

ω Maximum ration.

Ω_n Summation constant, where $n = +1, -3, +3, -1$.

Ω_0 Solid angle.

Glossary

Age group. The group of fish at a given age. 0 group fish are those in the first year of life; I group fish are those in the second year of life; II group fish are fish in their third year. A fish born on 1 April remains in the 0 group until 1 April in the subsequent year, after which it is allocated to the I group, II group, etc.

"Arctic city." A fleet of large trawlers from many nations that used to fish for cod in a small patch, usually in the Barents Sea.

Bathythermograph. An instrument which records temperature and depth continuously as it is lowered into the sea.

Beam trawl. A small trawl, the mouth of which is held open by a beam between two trawl heads or iron runners. Originally used by sailing smacks, it is used today on alternate sides of North Sea trawlers with powerful engines.

Cathode ray oscilloscope (CRT). An instrument that displays transient changes of voltage on a phosphorescent screen.

Chumming. A method of angling for albacore that gather in the shoal when live bait have been cast upon the surface of the sea.

Cohort. A group of fish, hatched in the same year, which grow together through their ages until it is extinguished; synonymous with "year class."

Cohort analysis. A series of catch equations at each age in a cohort, solved successively from the oldest age to the youngest, with estimated values of natural mortality and a "terminal" estimate of fishing mortality on the oldest age group, which is corrected by catches at age back from the oldest to the youngest age. It depends upon the ratio of catch in numbers at age at the end of one year to the stock in numbers at the beginning of the next.

Cod end. The long bag at the end of the trawl in which the catch of fish is collected; the size of mesh in the cod end is fixed to allow the smaller fish to escape and put on more weight before they are perhaps caught again.

Cran. Originally a measure of volume, equivalent to 37.5 imperial gallons. In weight there are 5.5. crans to a metric ton of herring.

Demersal. Fish living on, or close to, the seabed.

Discriminatory analysis. A method of discriminating between two populations, using an array of measured characters and minimizing the differences in the whole array.

Distance function. A measure of the "distance" between two populations in terms of the differences used either in discriminatory analysis or in genetic terms.

Drag. The resistance of a body to a fluid as it moves or is moved through it.

Drifter shot. A herring drifter shoots 70–95 nets at dusk and hauls them a few hours later or at the following dawn. The unit operation is termed a "drifter shot."

255

Drift net. Like a gill net, but the fleet of nets drives with the tide and with the drifter, the fishing boat that works the nets.

Ekman transport. The transport of water under wind stress at a right angle to the wind's direction, to the right in the northern hemisphere and to the left in the southern.

Electric fishing. The capture of fish with the use of an electric current in fresh water by which the fish swim automatically toward the anode.

El Niño. A southerly flow of transequatorial origin off the coast of Peru at Christmastime or just after, which occurs once every five or seven years on average. The flow lasts for six or eighteen months; while it persists the productivity characteristic of upwelling is reduced by a factor of three.

Eltonian pyramid. An ecosystem comprises four or five trophic levels, of which the lowest, of smallest animals, are the most abundant, and of which the highest, of largest animals, are the least abundant; the structure of trophic levels in numbers is called the Eltonian pyramid.

Elver. A young eel which has metamorphosed and which is usually found in estuaries or on its way up rivers.

Froude's law. "The work done by a body in a fluid varies as the product of drag and the square of the velocity."

Gill net. A curtain of netting suspended from the surface by floats. Single nets are linked into fleets of nets. Fish swim into them in darkness and are caught by their gill covers.

Grab. A spring-loaded double jaw which closes as it is dropped onto the bottom, enclosing a fixed area of soil on the seabed.

Hardy-Weinberg law. The distribution of two characters determined by a pair of genes where a_2 is the probability that the first appears in a phenotype and b_2 is the probability that the second does so: $(a_2^2 + 2a_2b_2 + b_2^2)$.

Heincke's law. "The size and age of the plaice in a definite part of the North Sea are inversely proportional to the density of their occurrence, but directly proportional to the distance of the locality from the coast and to its depth."

Hensen net. A vertically hauled plankton net with a restricting cone at the mouth of 70 cm in diameter; the mesh size is 60 meshes to the linear inch. The net was designed by Victor Hensen of Kiel toward the end of the nineteenth century.

Hjort maturity stage I. A virgin fish with no gonad growing in the body cavity.

Homing. The return of fish to their native spawning grounds.

Indicator. A conspicuous animal or plant, the appearance of which "indicates" an abnormal invasion of the area observed, by plants, animals, and, by inference, water.

Isopleth. Contour.

K line. An inverse relationship between growth efficiency and ration, developed by Paloheimo and Dickie.

Keeled scales. Scales on the ventral edge, or keel, of a herring, the numbers of which differ among spawning grounds and which are used as a meristic character.

Log normal. Where the logarithms of a set of variate value are distributed according to a normal curve. It is useful in biological material because the effects of many biological processes, like reproduction, appear in numbers as geometric series.

Longline. A line to which hooks are attached. A great longline of two or three miles in length is baited and laid on the seabed. A pelagic longline is suspended by floats from the surface, may or may not be baited, and may be as much as 80 km in length.

Marginal value. An increment of value.

Maximum sustainable yield (MSY). The greatest yield that can be taken from a fish stock year after year. Although it can be defined mathematically in a number of ways, it is perhaps best used today in a self-evident sense.

Meristic characters. Characteristics such as vertebral sum or a morphometric measurement which were formerly used in racial studies.

Metamorphosis. The change of shape that occurs when larvae become juvenile fish with all the fins and morphological characters of an adult.

Midwater. Any part of the water column between the surface and the seabed.

Migration. The movements of fish from feeding ground to spawning ground and back again, from nursery ground to feeding ground, and from spawning ground to nursery ground.

Model. A conceptual structure, or hypothesis, which may be tested by observation or experiment. Some models may be used experimentally, each trial being a conceptual experiment.

Natural mortality. The loss in numbers in a year class from one age group to the subsequent one, due to natural death.

Nursery ground. Where juvenile fish grow with least risk of predation; a beach or shallow bank.

Ogive. A cumulative percentage frequency distribution.

Orientation. The choice of a bearing as expressed in the direction of an animal's movement.

Otolith. Earstone, used by fish for its sense of balance. There is one in each plane of the semicircular canals on each side of the head, making six in all. Fisheries biologists use the biggest ones to determine the age of fishes.

Otter trawl. A large triangular bag of netting dragged along the seabed. The mouth is spread apart by two boards, one on each towing bridle.

Pair trawler. One of a pair of boats towing a trawl on or near the bottom or in the midwater.

Pelagic. Living in midwater.

Petersen's young-fish trawl. A small trawl made of shrimp netting, or of finer mesh, towed on a single bridle from which warps to spreading boards diverge and attach to the wings.

Plankton recorder. A robust net towed at up to 22 knots by merchant vessels. In it the plankton is caught and stored on a continuously moving band of nylon mesh.

Postage stamp plaice. The small juvenile plaice, the size of a postage stamp, found on the beaches where the nursery grounds are.

Pound net. A net on shore in which fish are impounded. They are also called set nets or fixed engines.

Purse seine. A curtain of net shot at the surface in the form of a circle; an encircling net. It is closed below by ropes passing through purse rings, and this action is called pursing.

Quota. The national share of an international Total allowable catch (TAC) estimated by scientists in a regional international body (for example, the International Council for the Exploration of the Sea).

Recruitment. The entrance of young fish of a year class into a fishery. The young fish recruit to a fishery over a period, sometimes less than a year and sometimes for two or three years.

Redd. A gravel bed in a river in which salmon lay their eggs.

Reynold's number. Ratio of mass forces to viscous forces in a fluid, which describes the transition from laminar to turbulent flow.

Salinity-temperature-depth recorder (STD). A probe which records salinity and temperature continuously as it descends in depth.

Scattering cross-section. The acoustic "area" of a target.

Siome. The edge of an oceanic boundary which is turbulent and noisy. The name is Japanese and describes the edge of the Kuroshio current.

Smolt. An adolescent salmon which has metamorphosed and which is found on its way downstream toward the sea.

Sonar. An apparatus that uses sound waves to detect objects underwater by measuring or classifying the echoes received from them. An echo sounder is a sonar that transmits vertically. In practice, a sonar is an apparatus other than an echo sounder, i.e., a sonar transmits horizontally.

Spent herring. A herring that has just spawned; a "shotten" herring.

Stanza. A distinct stage in growth, as, for example, the fresh water stage in the growth of any one of the Pacific salmon species.

Statistical rectangle. The ocean is divided into rectangles with sides of, for example, 30 International Nautical Miles, and catches are internationally classified by them.

Steady state. A population in a steady state may fluctuate about a mean, but does not increase or decline in a systematic way with time.

Swallow float. A metal float which remains at the same depth and transmits its position continuously as it drifts at that depth.

Swirl. A gyre on a small, temporary scale.

Tagging. The attachment of tags or marks to living fish, yielding information on migration or on growth and mortality between release and recapture.

Target strength. The ratio of received signal to transmitted signal from an object as at 1 m from the transmitter. It is expressed in watts/cm^2 or in db with reference to a standard 1 m sphere.

Total mortality. The loss in numbers in a year class from one age group to the subsequent one, due to all causes, including sometimes emigration.

Transponding acoustic tag. A small acoustic transmitter which is triggered by a signal. It is used in migration studies to record the movements of individual plaice with the Admiralty Research Laboratory sector scanner.

Upwelling. Under a longshore wind in subtropical seas, water is drifted offshore by the Ekman transport at the surface, and it is replaced by cooler water from below. Because the cool water is poor in plankton, the production cycle starts from very low levels as in temperate waters.

Whitebait. 0 group herring caught in estuaries, often mixed with sprats of about the same size.

Year brood or *Year class*. A brood of fish hatched in a given year. After the brood recruits to the fishery it reappears year after year until it is extinguished.

Zone electrophoresis. A method of separating proteins (for example, in blood sera) by their differential migration in an electric field.

References

Aasen, O., and A. Fridriksson. 1952. The Norwegian-Icelandic tagging experiments. Rit. Fiskideild. 1. 54 p.

Agger, P., I. Boetius, and H. Lassen. 1973. Error in the virtual population analysis. The effect of uncertainties in the natural mortality coefficient. J. Cons. Intern. Explor. Mer, 35:93.

Ahlstrom, E. H. 1966. Distribution and abundance of sardine and anchovy larvae in the California Current region off California and Baja California, 1951–64: A summary. U.S. Fish and Wildlife Serv. (Fisheries), Spec. Sci. Rep. 534. 71 p.

Allen, K. R. 1951. The Horokiwi Stream: A study of a trout population. Fish. Bull. N.Z. Mar. Dept., 10:1–231.

Allen, K. R. 1953. A method for computing the optimum size limit for a fishery. Nature, 172(4370):210.

Allen, K. R. 1969. Application of the Bertalanffy growth equation to problems of fisheries management: A review. J. Fish. Res. Board, Can., 26(9):2267–81.

Alm, G. 1959. Connection between maturity, size and age in fishes. Rep. Inst. Freshwater Res. Drottningholm, 40:5–145.

Alverson, D. L., and H. A. Larkins. 1969. Status of knowledge of the Pacific hake resources. Rep. Calif. Coop. Ocean. Fish. Invest., 13:24–31.

Alward, G. L. 1932. The sea fisheries of Great Britain and Ireland: A record of the development of the fishing industry and its world-wide ramifications. A. Gait, Grimsby. 549 p.

Ancellin, J. 1956. Le hareng du sud de la Mer du Nord et de la Manche orientale — observations de 1945 à 1954. Sci. et Pêche, 21:1–5.

Ancellin, J., and C. Nédelèc. 1959. Marquage de harengs en Mer du Nord et en Manche orientale (Campagne du "Président Théodore Tissier," Novembre 1957. Rev. Trav. Inst. Pêches Marit., 23:177–201.

Andersen, K. P., and E. Ursin. 1977. A multi-species extension of the Beverton and Holt theory of fishing, with accounts of phosphorus circulation and primary production. Meddr. Danm. Fisk. og Havunders., 7:319–435.

Andrewartha, H., and L. Birch. 1954. The distribution and abundance of animals. Univ. Chicago Press, Chicago. 782 p.

Backiel, T., and E. D. Le Cren. 1967. Some density relationships for fish population parameters, p. 261–93. In S. D. Gerking [ed.], The biological basis of freshwater fish production. Blackwell, Oxford. 495 p.

Bagenal, T. B. 1973. Fish fecundity and its relations with stock and recruitment. Rapp. Procès-Verb. Cons. Intern. Explor. Mer, 164:186–98.

Bailey, N. T. J. 1951. On estimating the size of mobile populations from recapture data. Biometrika, 38:293–306.

Bainbridge, R. 1960. Speed and stamina in three fish. J. Exp. Biol., 37:129–53.

Bannister, R. C. A., D. Harding, and S. J. Lockwood. 1974. Larval mortality and subsequent year-class strength in the plaice (*Pleuronectes platessa* L.), p. 21–37. *In* J. H. S. Blaxter [ed.], The early life history of fish. Springer-Verlag, Berlin. 765 p.

Baranov, F. I. 1918. On the question of the biological basis of fisheries. Nauchnyi issledovatelskii ikhtiologicheskii Institut Isvestia, 1(1):81–128.

Barets, A. 1961. Contribution à l'étude des systèmes moteurs "lents" et "rapides" du muscle latéral des téléostéans. Arch. Anat. Microsc. Morph. Exp., 50(1):91–187.

Batschelet, E. 1965. Statistical methods for the analysis of problems in animal orientation and certain biological rhythms. Amer. Inst. Biol. Sci. 55 p.

Battle, H. I. 1935. Digestion and digestive enzymes in the herring (*Clupea harengus* L.). J. Biol. Board, Can., 1(3):145–57.

Baxter, I. G. 1959. Fecundities of winter-spring and summer-autumn herring spawners. J. Cons. Intern. Explor. Mer, 25(1):73–80.

Bedford, B. C. 1966. English cod tagging experiments in the North Sea. Intern. Council Explor. Sea, Council Mtg., G 9. 9 p. Mimeograph.

Bertalanffy, L. von. 1934. Untersuchungen über die Gesetzlichkeit des Wachstums. I. Allgemeine Grundlagen der Theorie; mathematische und physiologische Gesetzlichkeiten des Wachstums bei Wassertieren. Arch. Entwicklungsmech., 131:613–52.

Bertelsen, E., and K. Popp Madsen. 1953–57. Young herring from the Bløden ground area. Ann. Biol., Cons. Intern. Explor. Mer, 9(1953):179–80; 10(1954):155–56; 12(1957):197–98.

Beverton, R. J. H. 1962. Long-term dynamics of certain North Sea fish populations, p. 242–64. *In* E. D. Le Cren and M. W. Holdgate [eds.], The exploitation of natural animal populations. Blackwell, London. 399 p.

Beverton, R. J. H. 1963. Maturation, growth, and mortality of clupeid and engraulid stocks in relation to fishing. Rapp. Procès-Verb. Cons. Intern. Explor. Mer, 154:44–67.

Beverton, R. J. H. 1964. Differential catchability of male and female plaice in the North Sea and its effect on estimates of stock abundance. Rapp. Procès-Verb. Cons. Intern. Explor. Mer, 155:103–112.

Beverton, R. J. H., and V. M. Hodder, eds. 1962. Report of working group of scientists on fishery assessment in relation to regulation problems. Suppl. Ann. Proc. Intern. Comm. N.W. Atlantic Fish., 11:81 p.

Beverton, R. J. H., and S. J. Holt. 1957. On the dynamics of exploited fish populations. H. M. Stationery Off., London, Fish. Invest., Ser. 2, Vol. 19. 533 p.

Beverton, R. J. H., and S. J. Holt. 1959. A review of the lifespans and mortality rates of fish in nature, and their relation to growth and other physiological characteristics, p. 142–80. *In* G. E. W. Wolstenholme and M. O'Connor [eds.], Ciba Foundation, Colloquia on ageing. V. The lifespan of animals. Churchill, London. 324 p.

Beverton, R. J. H., and S. J. Holt. 1964. Tables of yield functions for fishery assessment. F.A.O. Fish. Tech. Paper, No. 38. 49 p.

Beverton, R. J. H., and A. J. Lee. 1964. The influence of hydrographic and other factors on the distribution of cod on the Spitzbergen shelf. Intern. Comm. N.W. Atlantic Fish., Spec. Pub. No. 6:225–46.

Bidder, G. P. 1925. The mortality of plaice. Nature, 115(2892):495–96.

Birkett, L. 1969. The nitrogen balance in plaice, soles and perch. J. Exp. Biol. 50(2):375–86.

Bjerknes, J. 1961. "El Niño": A study based on analysis of ocean surface temperatures, 1935–57. Bull. Inter-Amer. Trop. Tuna Comm., 5(3):219–303.

Blaxter, J. H. S. 1965. The feeding of herring larvae and their ecology in relation to feeding. Rep. Calif. Coop. Ocean. Fish. Invest., 10:79–88.

Blaxter, J. H. S. 1968. Light intensity vision and feeding in young plaice. J. Exp. Mar. Biol. Ecol., 2(3):293–307.

Blaxter, J. H. S., and W. Dickson. 1959. Observations on the swimming speeds of fish. J. Cons. Intern. Explor. Mer, 24:472–79.

Blaxter, J. H. S., and G. Hempel. 1963. The influence of egg size on herring larvae (Clupea harengus L.). J. Cons. Intern. Explor. Mer, 28:211–40.

Bodenheimer, F. S. 1938. Problems of animal ecology. Oxford Univ. Press, London. 183 p.

Bodholt, H. 1969. Quantitative measurement of scattering layers. Simrad echo sounding systems, Bull. No. 3. 9 p.

Bodholt, H. 1971. Measuring target strength and back scattering strength. Simrad echo sounding systems, Bull. No. 5. 12 p.

Böhnecke, G. 1922. Salzegehalt und Strömungen der Nordsee. Veröffentlich.Inst. Meeresk. A. Geogr. Naturwiss. Reihe, No. 10:1–34.

Boiko, E. G. 1964. Evaluating natural mortality in the Azov zander (Stizostedion lucioperca). Fish. Res. Board, Can., Translation Series No. 541. 15 p.

Bolster, G. C. 1955. English tagging experiments. Rapp. Procès-Verb. Cons. Intern. Explor. Mer, 140(2):11–14.

Bolster, G. C., and J. P. Bridger. 1957. Nature of the spawning area of herrings. Nature, 179(4560):638.

Bostrøm, O. 1955. "Peder Ronnestad" Ekkolodding og meldetjeneste av Skreiforekomstene. Lofoten i tiden 1 March–2 Apr. 1955: Praktiske fiskeforsøk 1954 og 1955. Arsberet. Vedkomm. Norges Fisk., 9:66–70.

Brander, K. M. 1974. The effects of age-reading errors on the statistical reliability of marine fishery modelling, p. 181–91. In T. B. Bagenal [ed.], Ageing of fish. Proc. Int. Symp. Univ. Reading. Unwin Bros., Old Woking, Surrey, England.

Brander, K. M. 1977. The management of Irish Sea fisheries: A review. Min. Agr., Fish., Food, London, Lab. Leaflet No. 36. 40 p.

Brett, J. R. 1964. The respiratory metabolism and swimming performance of young sockeye salmon. J. Fish. Res. Board, Can., 21:1183–1226.

Brett, J. R. 1965. The relation of size to rate of oxygen consumption and sustained swimming speed of sockeye salmon (Oncorhynchus nerka). J. Fish. Res. Board, Can., 22:1491–1501.

Bridger, J. 1960. On the relationship between stock, larvae and recruits in the "Downs" herring. Cons. Intern. Explor. Mer, Herring Comm., Paper No. 159. 9 p. Mimeograph.

Brock, V., and R. Riffenburgh. 1960. Fish schooling: A possible factor in reducing predation. J. Cons. Intern. Explor. Mer, 25:307–17.

Brothers, E. G., C. P. Mathews, and R. Lasker. 1976. Daily growth increments in otoliths from larval and adult fishes. Fish. Bull. Nat. Mar. Fish. Serv. U.S., 74(1):1–8.

Brown, B. E., J. A. Brenhan, M. D. Grosslein, E. G. Heyerdahl, and R. C. Hennemuth. 1975. The effect of fishing on the marine finfish biomass in the Northwest Atlantic from the Gulf of Maine to Cape Hatteras. Res. Bull. Intern. Comm. N.W. Atlantic Fish., 12:49–68.

Brown, M. E. 1946a. The growth of brown trout (Salmo trutta Linn.) II. The growth of two-year-old trout at a constant temperature of 11.5° C. J. Exp. Biol., 22:130–44.

Brown, M. E. 1946b. The growth of brown trout (Salmo trutta Linn.). III. The effect of temperature on the growth of two-year-old trout. J. Exp. Biol., 22:145–55.

Bückmann, A. 1942. Die Untersuchungen der Biologischen Anstalt über die Ökologie der Heringsbrut in der südlichen Nordsee. Helgoland Wiss. Meeresunters., 3:1–57.

Bull, H. O. 1936. Studies on conditioned responses in fishes. VII. Temperature perception in teleosts. J. Mar. Biol. Ass., U.K., N.S., 21:1–27.

Burd, A. C. 1978. Long-term changes in North Sea herring stocks. Rapp. Procès-Verb. Cons. Intern. Explor. Mer, 172:137–53.

Burd, A. C., and J. Bracken. 1965. Studies on the Dunmore herring stock. I. A population assessment. J. Cons. Intern. Explor. Mer, 29(3):277–301.

Burd, A. C., and D. H. Cushing. 1962. I. Growth and recruitment in the herring of the southern North Sea. II. Recruitment to the North Sea herring stocks. H. M. Stationery Off., London, Fish. Invest., Ser. 2, Vol. 23(5). 71 p.

Burd, A. C., and W. Parnell. 1973. The relationship between larval abundance and stock in the North Sea herring. Rapp. Procès-Verb. Cons. Intern. Explor. Mer, 164:30–36.

Burgner, R. L., C. J. Dicostanzo, R. J. Ellis, G. Y. Harry, Jr., W. L. Hartman, O. E. Kerns, Jr., O. A. Mathisen, and W. F. Royce. 1969. Biological studies and estimates of optimum escapements of sockeye salmon in the major river systems in southwestern Alaska. Fish. Bull. U.S. Dept. Int., 67(2):405–59.

Burkenroad, M. 1948. Fluctuations in abundance of Pacific halibut. Peabody Mus. Natur. Hist., Yale Univ., New Haven, Conn., Bull. Bingham Oceanogr. Coll., 11(4). 81 p.

Carlisle, D. B., and E. J. Denton. 1959. On the metamorphosis of the visual pigments of *Anguilla anguilla* (L.). J. Mar. Biol. Ass., U.K., N.S., 38:97–102.

Cavalli-Sforza, L. L., and A. W. F. Edwards. 1967. Phytogenetic analysis: models and estimation procedures. Evolution, 21:550–70.

Chapman, D. G., R. J. Myhre, and G. M. Southward. 1962. Utilization of Pacific halibut stocks: Estimation of maximum sustainable yield 1960. Rep. Intern. Pacif. Halibut Comm., No. 31. 35 p.

Chapman, D. G., and D. S. Robson. 1960. The analysis of a catch curve. Biometrics, 16:354–68.

Clark, C. W. 1976. Mathematical bioeconomics: The optional management of renewable resources. Wiley Interscience, New York. 352 p.

Clark, F. N., and J. C. Marr. 1956. Population dynamics of the Pacific sardine. Calif. Coop. Ocean. Fish. Invest., Prog. Rep., 1 July 1953 to 31 March 1955, p. 11–48.

Clayden, A. D. 1972. Simulation of the changes in abundance of the cod (*Gadus morhua* L.) and the distribution of fishing in the North Atlantic. H. M. Stationery Off., London, Fish. Invest., Ser. 2, Vol. 27(1). 58 p.

Clutter, R. I., and L. E. Whitesel. 1956. Collection and interpretation of sockeye salmon scales. Bull. Intern. Pacif. Salmon Comm., No. 9. 159 p.

Colebrook, J. M. 1965. On the analysis of variation in the plankton, the environment and the fisheries. Intern. Comm. N.W. Atlantic Fish., Spec. Pub. No. 6:291–302.

Cooper, J. C., A. T. Scholz, R. M. Horrall, and A. D. Hasler. 1975. Experimental confirmation of the olfactory hypothesis with homing artificially imprinted coho salmon (*Oncorhynchus kisutch*). J. Fish. Res. Board, Can., 33:703–10.

Corlett, J. 1958. Distribution of larval cod in the western Barents Sea, p. 281–88. *In* Some problems for biological fishery survey and techniques for their solution. Intern. Comm. N.W. Atlantic Fish., Spec. Pub. No. 1.

Cormack, R. M. 1969. The statistics of capture-recapture methods. Oceanogr. Mar. Biol., 6:455–506.

Crutchfield, J. A. 1965. The fisheries: Problems in resource management. Univ. Washington Press, Seattle. 134 p.

Cushing, D. H. 1952. Echo-surveys of fish. J. Cons. Intern. Explor. Mer, 18:45–60.

Cushing, D. H. 1955. Production and a pelagic fishery. H. M. Stationery Off., London, Fish. Invest., Ser. 2, Vol. 18(7). 104 p.

Cushing, D. H. 1957. The number of pilchards in the Channel. H. M. Stationery Off., London, Fish. Invest., Ser. 2, Vol. 21(5). 27 p.

Cushing, D. H. 1959a. On the effect of fishing on the herring of the southern North Sea. J. Cons. Intern. Explor. Mer, 24(2):283–307.

Cushing, D. H. 1959b. The seasonal variation in oceanic production as a problem in population dynamics. J. Cons. Intern. Explor. Mer, 24(3):455–64.

Cushing, D. H. 1960. The East Anglian fishery in 1959. World Fishing, 9(8):51–58.

Cushing, D. H. 1961. On the failure of the Plymouth herring fishery. J. Mar. Biol. Ass., U.K., N.S., 41(3):799–816.

Cushing, D. H. 1964. The work of grazing in the sea, p. 207–25. In D. J. Crisp [ed.], Grazing in terrestrial and marine environments. Blackwell, London. 322 p.

Cushing, D. H. 1967. The grouping of herring populations. J. Mar. Biol. Ass., U.K., N.S., 47:193–208.

Cushing, D. H. 1968a. The Downs stock of herring during the period 1955–1966. J. Cons. Intern. Explor. Mer, 32(2):262–69.

Cushing, D. H. 1968b. Direct estimation of a fish population acoustically. J. Fish. Res. Board, Can., 25(11):2349–64.

Cushing, D. H. 1969a. The regularity of the spawning season of some fishes. J. Cons. Intern. Explor. Mer, 33(1):81–92.

Cushing, D. H. 1969b. The fluctuation of year classes and the regulation of fisheries. Fiskeridir. Skr. Ser. Havundersøk., 15:368–79.

Cushing, D. H. 1971a. Upwelling and the production of fish. Adv. Mar. Biol., 9:255–335.

Cushing, D. H. 1971b. The dependence of recruitment on parent stock in different groups of fishes. J. Cons. Intern. Explor. Mer, 33:340–62.

Cushing, D. H. 1972a. The production cycle and numbers of mature fish. Symp. Zool. Soc., London, 29:213–32.

Cushing, D. H. 1972b. A history of some of the international fisheries commissions. Proc. Roy. Soc. Edinburgh, Section B, 73:361–90.

Cushing, D. H. 1973a. Dependence of recruitment on parent stock. J. Fish. Res. Board, Can., 30:1965–76.

Cushing, D. H. 1973b. Detection of fish. Pergamon, New York. 200 p.

Cushing, D. H. 1974. The possible density dependence of larval mortality and adult mortality in fishes, p. 21–38. In J. H. S. Blaxter [ed.], The early life history of fish. Springer-Verlag, Berlin. 765 p.

Cushing, D. H. 1975a. The natural mortality of the plaice. J. Cons. Intern. Explor. Mer, 36(2):150–57.

Cushing, D. H. 1975b. Marine ecology and fisheries. Cambridge University Press, Cambridge. 277 p.

Cushing, D. H. 1976a. The impact of climatic change on fish stocks in the North Atlantic. Geogr. J., 142(2):216–27.

Cushing, D. H. 1976b. Biology of fishes in the pelagic community, p. 317–40. In D. H. Cushing and J. J. Walsh [eds.], The ecology of the seas. Blackwell, London. 467 p.

Cushing, D. H. 1977a. The Atlantic fisheries commissions. Mar. Pol., 1(3):230–38.

Cushing, D. H. 1977b. The problems of stock and recruitment, p. 116–35. In J. A. Gulland [ed.], Fish population dynamics. John Wiley, New York. Sydney. 369 p.

Cushing, D. H. 1978a. The present state of acoustic survey. J. Cons. Intern. Explor. Mer, 38(1):28–32.

Cushing, D. H. 1978b. The upper trophic levels in upwelling areas, p. 101–10. In R. Boje and M. Tomczak, [eds.], Upwelling ecosystems. Springer-Verlag, Berlin. 303 p.

Cushing, D. H. 1980. The decline of the herring stocks and the gadoid outburst. J. Cons. Intern. Explor. Mer, 39(1):74–85.

Cushing, D. H., and J. P. Bridger. 1966. The stock of herring in the North Sea and changes due to fishing. H. M. Stationery Off., London, Fish. Invest., Ser. 2, Vol. 25(1). 123 p.

Cushing, D. H., and A. C. Burd. 1957. On the herring of the southern North Sea. H. M. Stationery Off., London, Fish. Invest., Ser. 2, Vol. 20(11). 31 p.

Cushing, D. H., and R. R. Dickson. 1976. The biological response in the sea to climatic changes. Adv. Mar. Biol., 14:1–122.

Cushing, D. H., and J. G. K. Harris. 1973. Stock and recruitment and the problem of density dependence. Rapp. Procès-Verb. Cons. Intern. Explor. Mer, 164:142–55.

Cushing, D. H., and J. W. Horwood. 1977. Development of a model of stock and recruitment, p. 21–36. *In* J. H. Steele, [ed.], Fisheries mathematics. Academic Press, New York. 198 p.

Cushing, D. H., and H. F. Nicholson. 1963. Studies on a *Calanus* patch. IV. Nutrient salts off the north-east coast of England in the spring of 1954. J. Mar. Biol. Ass., U.K., N.S., 43(2):373–86.

Cushing, J. E. 1956. Observations on serology of tuna. U.S. Fish and Wildlife Serv., Spec. Sci. Rep. — Fish, No. 183. 14 p.

Cushing, J. E., and L. M. Sprague. 1953. Agglutinations of the erythrocytes of various fishes by human and other sera. Amer. Nat., 87(836):307–15.

Daan, N. 1973. A quantitative analysis of the food intake of North Sea cod, *Gadus morhua*. Neth. J. Sea Res., 6(4):479–517.

Daan, N. 1975. Consumption and production in North Sea cod (*Gadus morhua*): An assessment of the ecological status of the stock. Neth. J. Sea Res., 9(1):24–55.

Daan, N. 1978. Changes in cod stocks and cod fisheries in the North Sea. Rapp. Procès. Verb. Cons. Intern. Explor. Mer, 172:39–57.

Dahl, K. 1907. The scales of the herring as a means of determining age, growth and migration. Rep. Norweg. Fish. Mar. Invest., 2(6):1–36.

Dannevig, E. H. 1956. Cod populations identified by a chemical method. Fiskeridir. Skr. Ser. Havundersøk., 11 (6). 13 p.

Dannevig, G. 1954. The feeding grounds of the Lofoten cod. Rapp. Procès-Verb. Cons. Intern. Explor. Mer, 136:87–102.

Davies, D. H. 1957. The biology of the South African pilchard. Dept. Comm. Ind., Div. Fish., Union South Africa, Invest. Rep., No. 32:1–10.

Davies, D. H. 1958. The South African pilchard (*Sardinops ocellata*): The predation of sea birds in the commercial fishery. Dept. Comm. Ind., Div. Fish., Union South Africa, Invest. Rep., No. 31:1–15.

Deevey, E. S. 1947. Life tables for natural populations of animals. Quart. Rev. Biol., 22:283–314.

Devold, F. 1952. A contribution to the study of the migrations of the Atlanto-Scandian herring. Rapp. Procès-Verb. Cons. Intern. Explor. Mer, 131:103–7.

Devold, F. 1963. The life history of the Atlanto-Scandian herring. Rapp. Procès-Verb. Cons. Intern. Explor. Mer, 154:98–108.

Devold, F. 1966. Vintersildinnsigent 1966. Fiskets Gang, 52(16):299–301.

Dickie, L. M. 1963. Estimation of mortality rates of Gulf of St. Lawrence cod from results of a tagging experiment. Intern. Comm. N.W. Atlantic Fish., N. Atlantic Fish Marking Symp., Spec. Pub. No. 4:71–80.

Dickson, R. R., H. H. Lamb, S. A. Malmberg, and J. M. Colebrook. 1975. Climatic reversal in northern North Atlantic. Nature, 256(5517):479–81.

Dickson, R. R., J. G. Pope, and M. J. Holden. 1974. Environmental influence on the survival of North Sea cod, p. 69–80. *In* J. H. S. Blaxter [ed.], The early life history of fish. Springer-Verlag, Berlin. 765 p.

Dietrich, G. 1954. Verteilung, Ausbreitung und Vermischung der Wasserkörper in der süd-westlichen Nordsee auf Grund der Ergebnisse der "Gauss" — Fahrt im Februar–März 1952, Ber.deutsch.wiss.Komm Meeresf., NS, 13(2):104–29.

Dietrich, G. 1965. New hydrographical aspects of the Northwest Atlantic. Intern. Comm. N.W. Atlantic Fish., Spec. Pub. No. 6:29–51.

Drilhon, A., and J. M. Fine. 1969. Les groupes des transferrines dans le genre *Anguilla* L. Rapp. Procès-Verb. Cons. Intern. Explor. Mer, 161:122–25.

Edwards, R. R. C., D. M. Finlayson, and J. H. Steele. 1969. The ecology of 0-group place and common dabs in Loch Ewe. II. Experimental studies of metabolism. J. Exp. Mar. Biol. Ecol., 3:1–17.

Ege, V. 1939. A revision of the genus *Anguilla* Shaw, a systematic, phylogenetic and geographical study. Dana Rep., 3(16):1–256.

Eggvin, J. 1964. Water movement in the central part of the Norwegian Sea based on recent material. Intern. Council Explor. Sea, Council Mtg., 138. 12 p. Mimeograph.

Ehrenbaum, E., and H. Marukawa. 1913. Über Altersbestimmung und Wachstum beim Aal. Z. Fisch. Hilfswissenschaft., 14:89–127.

English, T. S. 1964. A theoretical model for estimating the abundance of planktonic fish eggs. Rapp. Procès-Verb. Cons. Intern. Explor. Mer, 155:174–82.

Farris, D. A. 1960. The effect of three different types of growth curves on estimates of larval fish survival. J. Cons. Intern. Explor. Mer, 25(3):294–306.

Fine, J. M., A. Drilhon, G. Ridgeway, and G. A. Boffin. 1967. Les groupes des transferrines dans le genre *Anguilla:* Différences phénotypiques des transferrines chez *Anguilla anguilla* et *Anguilla rostrata*. Comptes Rendues, Acad. Sci., Paris, 265:58–60.

Fisher, R. A. 1936. The use of multiple measurements in taxonomic problems. Ann. Eug., London, 7:178–88.

Flores, L. A. 1967. Informe preliminar del Crucero 6611 de la primavera de 1966 (Cabo Blanco-Punta Coles). Inf. Inst. Mar. Peru, Informe No. 17. 16 p.

Foerster, R. E. 1936. The return from the sea of sockeye salmon (*Oncorhynchus nerka*) with special reference to percentage survival, sex proportions and progress of migration. J. Biol. Board, Can., 3:26–42.

Foerster, R. E. 1968. The sockeye salmon (*Oncorhynchus nerka*). Bull. Fish. Res. Board, Can. 162 p.

Fox, W. W. 1970. An experimental surplus-yield model for optimizing exploited fish populations. Trans. Amer. Fish. Soc., 99(1):80–88.

Fridriksson, A. 1934. On the calculation of age-distribution within a stock of cod by means of relatively few age-determinations as a key to measurements on a large scale. Rapp. Procès-Verb. Cons. Intern. Explor. Mer, 86:1–5.

Fujino, K. 1970. Immunological and biochemical genetics of tunas. Trans. Amer. Fish. Soc., 99(1):152–78.

Fukuda, Y. 1962. On the stocks of halibut and their fisheries in the Northeast Pacific. Intern. N. Pacific Fish. Comm., Bull. No. 7:39–50.

Fukuhara, F. M., S. Murai, J.-J. Lalanne, and A. Sribhibhadh. 1962. Continental origin of red salmon as determined from morphological characters. Intern. N. Pacific Fish. Comm., Bull. No. 8:15–109.

Fulton, T. W. 1897. On the growth and maturation of the ovarian eggs of teleostean fishes. 16th Ann. Rep., Fish. Board, Scot., 3:88–124.

Garrod, D. J. 1963. An estimation of the mortality rates in a population of *Tilapia esculenta* Graham (Pisces, Cichlidae) in Lake Victoria, East Africa. J. Fish. Res. Board, Can., 20:195–227.

Garrod, D. J. 1967. Population dynamics of the Arcto-Norwegian cod. J. Fish. Res. Board, Can., 24(1):145–90.

Garrod, D. J. 1969. Empirical assessments of catch/effort relationships in North Atlantic cod stocks. Res. Bull., Intern. Comm. N.W. Atlantic Fish., 6:26–34.

Garrod, D. J., and B. W. Jones. 1974. Stock and recruitment relationship in the Northeast Arctic cod stock and the implications for management of the stock. J. Cons. Intern. Explor. Mer, 36(1):35–41.

Garstang, W. 1900–1903. The impoverishment of the sea. J. Mar. Biol. Ass., U.K., N.S., 6:1–70.

Gilbert, C. H. 1914. Contributions to the life history of the sockeye salmon. No. 1. Rep. Comm. Fish., B.C. [1913], p. 53–78.

Gilis, C. 1957. Evolution dans le temps et dans l'espace de la composition des concentrations de harengs exploitées par les pêcheurs belges dans la Mer du Nord au cours de la période 1951–1955. Rapp. Procès-Verb. Cons. Intern. Explor. Mer, 143(1):34–42.

Glover, R. S., G. A. Robinson, and J. M. Colebrook. 1972. Plankton in the North Atlantic: An example of the problems of analyzing variability in the environment, p. 439–45. *In* M. Ruivo [ed.], marine pollution and sea life. Fishing News (Books), London. 624 p.

Goddard, G. C., and V. G. Welsby. 1977. Statistical measurements of the acoustic target strength of live fish. Rapp. Procès-Verb. Cons. Intern. Explor. Mer, 170:70–73.

Gompertz, B. 1825. On the nature of the functions expressive of the law of human mortality, and on a new mode of determining the value of life contingencies. Phil. Trans. Roy. Soc., 115:513–85.

Graham, J. J., S. B. Chenoweth, and C. W. Davis. 1972. Abundance, distribution, movements and lengths of larval herring along the western coast of the Gulf of Maine. Fish. Bull., 70(2):307–21.

Graham, M. 1933. Report on the North Sea cod. H. M. Stationery Off., London, Fish Invest., Ser. 2, Vol. 13(4). 160 p.

Graham, M. 1935. Modern theory of exploiting a fishery, and application to North Sea trawling. J. Cons. Intern. Explor. Mer, 10(2):264–74.

Graham, M. 1938. Rates of fishing and natural mortality from the data of marking experiments. J. Cons. Intern. Explor. Mer, 10:264–74.

Graham, M. 1958. Fish population assessment by inspection, p. 67–68. *In* Some problems for biological fishery survey and techniques for their solution. Intern. Comm. N.W. Atlantic Fish., Spec. Pub. No. 1.

Grassi, G. B. 1896. The reproduction and metamorphosis of the common eel (*Anguilla vulgaris*). Quart. J. Microscop. Sci., 39:371–85.

Gray, J. 1926. The kinetics of growth. J. Exp. Biol., 6(3):248–74.

Gray, J. 1968. Animal locomotion. Weidenfeld and Nicolson, London. 479 p.

Great Britain. Ministry of Agriculture, Fisheries, and Food, U.K. 1885–1965. Sea fisheries statistical tables. H. M. Stationery Off., London. 80 vol.

Great Britain. Ministry of Agriculture, Fisheries, and Food, and Department of Agriculture and Fisheries, Scotland, 1962. Fish stock record, 1961. Mar. Lab., Aberdeen, and Fish Lab., Lowestoft. 52 p.

Greer Walker, M. 1970. Growth and development of the skeletal muscle fibres of the cod (*Gadus morhua* L.). J. Cons. Intern. Explor. Mer, 33(2):228–44.

Grosslein, M. D., R. W. Langton, and M. P. Sissenwine. In press. Recent fluctuations in Pelagic fish stocks of the Northwest Atlantic, Georges Bank region, in relationship to species interactions. Cons. Intern. Explor. Mer, Symp., Biol. basis of Pelagic fish stock management, 25. 52 p.

Guillen, O. 1976. El sistema de la corriente Peruana. I. Aspectos fisicos. I.D.O.E. Workshop on El Niño, Guayaquil. F.A.O. Fish. Rep., 185:243–84.

Gulland, J. A. 1955a. Estimation of growth and mortality in commercially exploited fish populations. H. M. Stationery Off., London, Fish Invest., Ser. 2, Vol. 18(9). 46 p.

Gulland, J. A. 1955b. On the estimation of population parameters from marked members. Biometrika, 42:269–70.

Gulland, J. A. 1961. Fishing and the stocks of fish at Iceland. H. M. Stationery Off., London, Fish Invest., Ser. 2, Vol. 23(4). 52 p.

Gulland, J. A. 1963. The estimation of fishing mortality from tagging experiments. Intern. Comm. N.W. Atlantic Fish., N. Atlantic Fish Marking Symp., Spec. Pub. No. 4:218–27.

Gulland, J. A. 1964. On the measurement of abundance of fish stocks. Symposium held in Madrid, 25–28 Sept. 1963. Rapp. Procès-Verb. Cons. Intern. Explor. Mer, 155. 223 p.

Gulland, J. A. 1965. Estimation of mortality rates. Annex Northeast Arctic Working Group. Council Meeting. G 3. 9 p. Mimeograph.

Gulland, J. A. 1966. North Sea plaice stocks. Min. Agr., Fish., Food, London, Lab. Leaflet No. 11. 18 p.

Gulland, J. A. 1968a. The concept of the marginal yield from exploited fish stocks. J. Cons. Intern. Explor. Mer, 32(2):256–61.

Gulland, J. A. 1968b. The concept of the maximum sustainable yield and fishery management. F.A.O. Tech. Rep., No. 50.

Gulland, J. A. 1969. Manual of methods for fish stock assessment. Part I. Fish population analyses. F.A.O. Manuals in Fisheries Science, No. 4. 154 p.

Gulland, J. A. 1972. Population dynamics of world fisheries. Div. Mar. Res., Univ. Washington, Sea Grant Program, Seattle. 336 p.

Gulland, J. A. 1977. The analysis of data and development of models, p. 67–95. In J. A. Gulland [ed.], Fish population dynamics. John Wiley, New York. 372 p.

Gulland, J. A., and G. R. Williamson. 1962. Transatlantic journey of a tagged cod. Nature, 195(4844):921.

Haldane, J. B. S. 1953. Animal populations and their regulation. New Biol., 15:9–24.

Halliday, R. G. 1971. Recent events in the haddock fishery of the eastern Scotian shelf. Res. Bull. Intern. Comm. N.W. Atlantic Fish., 8:49–58.

Hansen, P. M., A. S. Jensen, and A. V. Tåning. 1935. Cod marking experiments in the waters of Greenland, 1924–33. Meddr. Danm. Fisk. og Havunders., 10(1). 119 p.

Harding, D., J. H. Nichols, and D. Tungate. 1978. The spawning of the plaice (Pleuronectes platessa L.) in the southern North Sea and English Channel. Rapp. Procès-Verb. Cons. Intern. Explor. Mer, 172:102–13.

Harding, D., and J. W. Talbot. 1973. Recent studies on the eggs and larvae of the plaice (Pleuronectes platessa L.) in the Southern Bight. Rapp. Procès-Verb. Cons. Intern. Explor. Mer, 164:261–69.

Hardy, A. C., G. T. D. Henderson, C. E. Lucas, and J. H. Fraser. 1936. The ecological relations between the herring and the plankton investigated with the plankton indicator. J. Mar. Biol. Ass., U.K., N.S., 21:147–304.

Harris, J. G. K. 1975. The effect of density-dependent mortality on the shape of the stock and recruitment curve. J. Cons. Intern. Explor. Mer, 36(2):144–49.

Hart, T. J., and R. I. Currie. 1960. The Benguela current. Discovery Reports, 31:1–297.

Hasler, A. D. 1966. Underwater guideposts: Homing of salmon. Univ. Wisconsin Press, Madison. 155 p.

Haug, A., and O. Nakken. 1977. Echo abundance indices of 0-group fish in the Barents Sea, 1965–1972. Rapp. Procès-Verb. Cons. Intern. Explor. Mer, 170:259–64.

Heincke, F. 1898. Naturgeschichte des Herings. Arch. Deutsch Seefisch., 2:128–223.

Heincke, F. 1913. Untersuchungen über die Scholle. Generalbericht. I. Schollenfischerei und Schonmassregeln. Vorläufige kurze Übersicht über die wichtigsten Ergebnisse des Berichts. Rapp. Procès-Verb. Cons. Intern. Explor. Mer, 16:1–70.

Hempel, G. 1955. Zur Beziehung zwischen Bestandsdichte und Wachstum in der Schollenbevölkerung der deutschen Bucht. Ber. dt. Wiss. Kommn. Meeresforsch., 15(2):132–44.

Hempel, G., ed. 1978. North Sea fish stocks — recent changes and their causes. Rapp. Procès-Verb. Cons. Intern. Explor. Mer, 172. 449 p.

Henry, K. A. 1961. Racial identification of Fraser River sockeye salmon by means of scales and its applications to salmon management. Bull. Intern. Pacif. Salmon Comm., No. 12. 47 p.

Hickling, C. F. 1931. The structure of the otolith of the hake. Quart. J. Microscop. Sci., 74:547–61.

Hildemann, W. H. 1956. Goldfish erythrocyte antigens and serology. Science, 124:315–16.

Hjort, J. 1914. Fluctuations in the great fisheries of northern Europe viewed in the light of biological research. Rapp. Procès-Verb. Cons. Intern. Explor. Mer, 20:1–228.

Hjort, J. 1926. Fluctuations in the year classes of important food fishes. J. Cons. Intern. Explor. Mer, 1:1–38.

Hodgson, W. C. 1925. Investigations into the age, length and maturity of the herring of the southern North Sea. II. The composition of the catches in 1922–1924. H. M. Stationery Off., London, Fish. Invest., Ser. 2, Vol. 8(5). 48 p.

Hodgson, W. C. 1957. The herring and its fishery. Routledge and Kegan Paul, London. 197 p.

Höglund, H. 1955. Swedish herring tagging experiments 1949–1953. Rapp. Procès-Verb. Cons. Intern. Explor. Mer, 140(2):19–29.

Holden, M. J. 1973. Are long-term sustainable fisheries for elasmobranchs possible? Rapp. Procès-Verb. Cons. Intern. Explor. Mer, 164:360–67.

Holling, C. S. 1965. The functional response of predators to prey density and its role in mimicry and population regulation. Mem. Entom. Soc. Can., No. 45. 60 p.

Holling, C. S. 1973. Resilience and stability of ecological systems. Ann. Rev. Ecol. Syst., 4:1–24.

Holt, S. J., and P. H. Thomas. 1957. A mathematical model for line and gill-net fisheries. Intern. Comm. N.W. Atlantic Fish., Spec. Pub. No. 19. 15 p.

Horwood, J. W. 1975. Interactive fisheries: A two species Schaefer model. Selected papers 1976. Intern. Comm. N.W. Atlantic Fish., 1:151–55.

Horwood, J. W. 1976. On the joint exploitation of krill and whales. Intern. Council Explor. Sea, Council Mtg., N12. 6 p.

Houvenaghel, G. T. 1978. Oceanographic conditions in the Galapagos Archipelago and their relationships with life on the islands, p. 181–200. In R. Boje and M. Tomczak [eds.], Upwelling ecosystems. Springer-Verlag, Berlin. 303 p.

Howard, L. O., and W. F. Fiske. 1911. The importation into the United States of the parasites of the gipsy moth and the brown tail moth. U.S. Dept. Agr. Bur. Ent. Bull., 91:1–312.

Hubbs, C. L. 1921. An ecological study of the life history of the fresh-water atherine fish Labidesthes sicculus. Ecology, 2:262–76.

Hunter, J. R. 1972. Swimming and feeding behaviour of larval anchovy Engraulis mordax. Fish. Bull., 70(3):821–38.

Hunter, J. R., and G. L. Thomas. 1974. Effect of prey distribution and density on the searching and feeding behaviour of larval anchovy Engraulis mordax, p. 559–74. In J. H. S. Blaxter [ed.], The early life history of fish. Springer-Verlag, Berlin. 765 p.

Hylen, A., L. Midttun, and G. Saetersdal. 1961. Torskeundersøkelsene i Lofoten og i Barentshavet 1960. Fisken og Havet., No. 2:1–14.

Iles, T. D. 1967. Growth studies on North Sea herring. I. The second year's growth (I-group) of East Anglian herring, 1939–63. J. Cons. Intern. Explor. Mer, 31:56–76.

Iles, T. D. 1968. Growth studies on North Sea herring. II. 0-group growth of East Anglian herring. J. Cons. Intern. Explor. Mer, 32(1):98–116.

Iles, T. D. 1973. Interaction of environment and parent stock size in determining recruitment in the Pacific sardine as revealed by analysis of density-dependent 0-group growth. Rapp. Procès-Verb. Cons. Intern. Explor. Mer, 164:228–40.

International Commission on Whaling. 1964. Fourteenth report of the commission. Intern. Whaling Comm., London. 122 p.

International Decade of Oceanic Exploration. 1976. Report of the workshop on the phenomenon known as El Niño, Guayaquil, Ecuador, 4–12 December 1974. F.A.O. Fish. Rep. No. 163. 24 p.

International North Pacific Fisheries Commission. 1966. Annual report. Vancouver. 127 p.

Iselin, C. O. 1936. A study of the circulation of the western North Atlantic. Papers Phys. Oceanogr., 4(4):1–97.

Ivlev, V. S. 1961. The experimental ecology of the feeding of fishes. Yale Univ. Press, New Haven. 302 p.

Iwata, M. 1975. Genetic identification of wall-eye pollock (Theragra chalcogramma) populations on the basis of tetrazolium oxidaze polymorphism. Comp. Biochem. Physiol., 50(B):197–201.

Jamieson, A., and B. W. Jones. 1967. Two races of cod at Faroe. Heredity, 22(4):610–12.

Jamieson, A., and J. Jonsson. 1971. The Greenland component of spawning cod at Iceland. Rapp. Procès-Verb. Cons. Intern. Explor. Mer, 161:65–72.

Jamieson, A., and R. J. Turner. 1979. The extended series of Tf alleles in Atlantic cod (*Gadus morhua* L.), p. 699–729. *In* B. Battaglia and J. Beardmore [eds.], Marine organisms: Genetics, ecology, and evolution. Plenum Press, New York.

Johannesson, K. A., and G. F. Losse. 1977. Methodology of acoustic estimations of fish abundance in some UNDP/FAO resource survey projects. Rapp. Procès-Verb. Cons. Intern. Explor. Mer, 170:296–318.

Johnson, M. S. 1971. Adaptive lactate dehydrogenase variation in the crested blenny, *Anoplarchus*. Heredity, 27:205–26.

Johnson, W. E. 1965. On mechanisms of self regulation of population abundance in *Oncorhyncus nerka*. Mitt. Intern. Verein Limnol., 13:66–87.

Jolly, G. M. 1965. Explicit estimates from capture-recapture data with both death and immigration: stochastic model. Biometrika, 52(1/2):225–47.

Jones, F. R. Harden. 1963. The reaction of fish to moving backgrounds. J. Exp. Biol., 40(3):437–46.

Jones, F. R. Harden. 1968. Fish migration. Arnold, London. 325 p.

Jones, F. R. Harden, G. P. Arnold, M. Greer Walker, and P. Scholes. 1979. Selective tidal stream transport and the migration of plaice (*Pleuronectes platessa* L.) in the southern North Sea. J. Cons. Intern. Explor. Mer, 38(1):331–37.

Jones, F. R. Harden, M. Greer Walker, and G. P. Arnold. 1978. Tactics of fish movement in relation to migration strategy and water circulation, p. 185–207. *In* H. Charnock and Sir George Deacon [eds.], Advances in Oceanography. Plenum Press, New York. 356 p.

Jones, F. R. Harden, A. R. Margetts, M. Greer Walker, and G. P. Arnold. 1977. The efficiency of the Granton otter trawl determined by sector-scanning sonar and acoustic transponding tags. Rapp. Procès-Verb. Cons. Intern. Explor. Mer, 170:45–51.

Jones, F. R. Harden, and P. Scholes. 1974. The effect of low temperature on cod (*Gadus morhua* L.). J. Cons. Intern. Explor. Mer, 35(3):258–71.

Jones, J. W. 1959. The salmon. Collins, London, 192 p.

Jones, R. 1956. The analysis of trawl haul statistics with particular reference to the estimation of survival rates. Rapp. Procès-Verb. Cons. Intern. Explor. Mer, 140(1):30–39.

Jones, R. 1959. A method of analysis of some tagged haddock returns. J. Cons. Intern. Explor. Mer, 25:58–72.

Jones, R. 1961. The assessment of the long term effects of changes in gear selectivity and fishing effort. Mar. Res., 2. 19 p.

Jones, R. 1964. Estimating population size from commercial statistics when fishing mortality varies with age. Rapp. Procès-Verb. Cons. Intern. Explor. Mer, 155:210–14.

Jones, R. 1973a. The stock and recruitment relation as applied to the North Sea haddock. Rapp. Procès-Verb. Cons. Intern. Explor. Mer, 164:156–73.

Jones, R. 1973b. Density dependent regulation of the numbers of cod and haddock. Rapp. Procès-Verb. Cons. Intern. Explor. Mer, 164:156–73.

Jones, R. 1976a. Growth of fishes, p. 251–79. *In* D. H. Cushing and J. J. Walsh [eds.], The ecology of the sea. Blackwell, London. 467 p.

Jones, R. 1976b. The use of marking data in fish population analysis. F.A.O. Fish. Tech. Paper, No. 153. 42 p.

Jones, R., and W. B. Hall. 1973. A simulation model for studying the population dynamics of some fish species, p. 35–59. *In* M. S. Bartlett and R. W. Hiorns [eds.], The mathematical theory of the dynamics of biological populations. Academic Press, New York. 347 p.

Jordan, R. 1976. Biologia de la anchoveta. I. Resumen del conocimiento actual. I.D.O.E. Workshop on El Niño, Guayaquil. F.A.O. Fish. Rep., 185:359–99.

Jordan, R., and A. C. de Vildoso. 1965. La anchoveta (*Engraulis ringens* J.). Conocimiento actual sobre su biología, ecología, y pesquería. Inst. Mar Peru, Informe 6:1–52.

Kerr, S. R. 1971a. A simulation model of lake trout growth. J. Fish. Res. Board, Can., 28:815–19.

Kerr, S. R. 1971b. Analysis of laboratory experimentation growth efficiency of fishes. J. Fish. Res. Board, Can., 28:801–8.

Kerr, S. R. 1971c. Prediction of fish growth efficiency in nature. J. Fish. Res. Board, Can., 28:809–14.

Kimura, D. K. 1977. Statistical assessment of the age-length key. J. Fish. Res. Board, Can., 34(3):317–24.

King, J. E., and T. S. Hida. 1957. Zooplankton abundance in the central Pacific. II. U.S. Fish and Wildlife Serv., Fish. Bull., 57(118):365–95.

Knauss, J. A. 1969. The transport of the Gulf Stream. Deep Sea Res. Suppl., 16:117–23.

Krefft, G. 1954. Untersuchungen zur Rassenfrage beim Heringe. Mitt. Inst. Seefischerei, Hamburg, 6:12–33.

Kvinge, T., A. J. Lee, and R. Saetre. 1968. Report on study of variability in the Norwegian Sea. Univ. Bergen Geofys. Inst. 31 p., figs.

Larkin, P. A., and A. S. Hourston. 1964. A model for simulation of the population biology of Pacific salmon. J. Fish. Res. Board, Can., 21(5):1245–65.

Larkin, P. A., and J. G. Macdonald. 1968. Factors in the population biology of the sockeye salmon of the Skeena River. J. Anim. Ecol., 37:229–58.

Larkin, P. A., R. A. Raleigh, and N. J. Wilimovsky. 1964. Some alternative premises for constructing theoretical reproduction curves. J. Fish. Res. Board, Can., 21(3):477–84.

Lasker, R. 1975. Field criteria for survival of anchovy larvae; the relation between inshore chlorophyll maximum layers and successful first feeding. Fish. Bull. Nat. Mar. Fish. Serv. U.S., 73(3):453–62.

Laurs, R. M., H. S. H. Yuen, and J. H. Johnson. 1977. Small-scale movements of albacore, *Thunnus alalunga*, in relation to ocean features as indicated by ultrasonic tracking and oceanographic sampling. Fish. Bull. Nat. Mar. Fish. Serv. U.S., 75(2):347–55.

Lea, E. 1929. The herring's scale as a certificate of origin. Its applicability to race investigations. Rapp. Procès-Verb. Cons. Intern. Explor. Mer, 54:21–34.

Lea, E. 1930. Mortality in the tribe of Norwegian herring. Rapp. Procès-Verb. Cons. Intern. Explor. Mer, 65:100–117.

Le Cren, E. D. 1958. Observations on the growth of perch (*Perca fluviatilis* L.) over twenty-two years with special reference to the effects of temperature and changes in population density. J. Anim. Ecol., 27:287–334.

Le Cren, E. D. 1962. The efficiency of reproduction and recruitment in freshwater fish, p. 283–302. *In* E. D. Le Cren and M. W. Holdgate [eds.], The exploitation of natural animal populations. Blackwell, London. 399 p.

Le Cren, E. D. 1965. Some factors regulating the size of populations of freshwater fish. Mitt. Intern. Verein Limnol., 13:88–105.

Lee, A. J. 1952. The influence of hydrography on the Bear Island cod fishery. Rapp. Procès-Verb. Cons. Intern. Explor. Mer, 131:74–102.

Le Gall, J. 1935. Le hareng, *Clupea harengus*, Linné. I. Les populations de l'Atlantique Nord Est. Ann. Inst. Océanogr., 15:1–215.

Legendre, R. 1934. La faune pélagique de l'Atlantique au large du Golfe de Gascogne recueillie dans des estomacs des germons. I. Poissons. Ann. Inst. Océanogr., 14(6):247–418.

Leslie, P. H. 1952. The estimation of population parameters from data obtained by means of the capture-recapture method. II. The estimation of total numbers. Biometrika, 39:363–88.

Lett, P. F., and A. C. Kohler. 1976. Recruitment: A problem of multispecies interaction and environmental perturbations with special reference to Gulf of St. Lawrence herring (*Clupea harengus harengus*). J. Fish. Res. Board, Can., 33:1353–71.

Li, C. C. 1955. Population genetics. Univ. Chicago Press, Chicago. 366 p.

Lighthill, M. J. 1971. Large-amplitude elongated-body theory of fish locomotion. Proc. Roy. Soc. Lond. B ser., 179:125–38.

Ligney, W., de. 1969. Serological and biochemical studies on fish populations. Oceanogr. Mar. Biol., 7:411–513.

Ljungman, A. 1882. Contribution towards solving the question of the secular periodicity of the great herring fisheries. U.S. Comm. Fish and Fisher., 7(7):497–503.

Lockwood, S. J. 1978. The mackerel: A problem in fish stock assessment. Min. Agr., Fish., Food, Direct. Fish. Res., Lowestoft, Lab. Leaflet No. 44. 18 p.

Lockwood, S. J. In press. Density-dependent mortality in 0-group plaice (*Pleuronectes platessa* L.) populations. J. Cons. Intern. Explor. Mer.

Lockwood, S. J., J. H. Nichols, and Wendy A. Dawson. In press. The estimation of a mackerel (*Scomber scombrus* L.) spawning stock size by plankton survey. J. Plankt. Res.

Lundbeck, J. 1954. German market investigations on cod, mainly in the northeastern area. Rapp. Procès-Verb. Cons. Intern. Explor. Mer, 136:33–39.

McAllister, C. D., T. R. Parsons, and J. D. H. Strickland. 1960. Primary production and fertility at station "P" in the north-east Pacific Ocean. J. Cons. Intern. Explor. Mer, 25(3):240–59.

McCall, A. D. 1974. The mortality rate of *Engraulis mordax* in southern California. Rep. Calif. Coop. Ocean. Fish. Invest., 17:131–35.

MacCall, A. D. 1976. Density dependence of catchability coefficient in the California Pacific sardine, *Sardinops sagax caerulea*, purse seine fishery. Rep. Calif. Coop. Ocean. Fish. Invest., 18:136–48.

Macer, C. T. 1967. The food web in the Red Wharf Bay (North Wales) with particular reference to young plaice (*Pleuronectes platessa*). Helgolander Meeresforsch, 15:560–73.

Macer, C. T. 1974. Some observations on the fecundity of the pilchard (*Sardina pilchardus* Walbaum) off the south-west of England. Intern. Council Explor. Sea, Council Mtg., J9. 6 p. Mimeograph.

Manzer, J. I., T. Ishida, A. E. Peterson, and M. G. Hanavan. 1965. Salmon of the North Pacific Ocean. Intern. N. Pacific Fish. Comm., Bull. No. 15. 452 p.

Margolis, L., F. C. Cleaver, Y. Fukuda, and H. Godfrey. 1966. Salmon of the North Pacific. VI. Sockeye salmon in offshore waters. Intern. N. Pacific Fish. Comm., Bull. No. 20:1–68.

Marliave, J. B. 1975. Seasonal shifts in the spawning site of a north east Pacific intertidal fish. J. Fish. Res. Board, Can., 32:1687–91.

Marr, J. C. 1951. On the use of the terms *abundance, availability* and *apparent abundance* in fishery biology. Copeia, 2:163–69.

Marr, J. C. 1956. The "critical period" in the early life history of marine fishes. J. Cons. Intern. Explor. Mer, 21(2):160–70.

Marr, J. C. 1957. The problem of defining and recognizing subpopulations of fishes. Spec. Sci. Rep., U.S. Dept. Int., Fish. and Wildlife Serv., 208:1–6.

Marr, J. C. 1960. The causes of major variations in the catch of the Pacific sardine *Sardinops caerulea* (Girard). World Sci. Meet. Biol. Sardines and Related Species, Proc., 3:667–791.

Marshall, P. T. 1957. Primary production in the Arctic. J. Cons. Intern. Explor. Mer, 23:173–77.

Martyshev, F. G. 1964. Effects of age of parents on carp fry. Dokl. mosk. sel.-khoz. Akad. K. A. Timirazeva. 95 p.

Maslov, N. A. 1944. Bottom fishes of the Barents Sea. Knipovich Inst., Murmansk, Trans., 8:3–186.

Mathisen, O. 1969. Growth of sockeye salmon in relation to abundance in the Kvichak district, Bristol Bay, Alaska. Fiskeridir. Skr. Ser. Havunders, 15(3):172–85.

Mathisen, O. A., T. R. Croker, and E. P. Nunnallee. 1977. Acoustic estimation of juvenile sockeye salmon. Rapp. Procès-Verb. Cons. Intern. Explor. Mer, 170:279–86.

Matsumoto, W. 1966. Distribution and abundance of tuna larvae in the Pacific Ocean, p. 221–30. *In* T. A. Manier [ed.], Proc. Gov. Conf. Centr. Pac. Res. Honolulu, State of Hawaii.

May, R. M. 1976. Models for single populations," p. 4–25. *In* R. M. May [ed.], Theoretical ecology: Principles and applications. Blackwell, London. 317 p.

Medawar, P. B., ed. 1945. Size, shape and age: Essays on growth and form presented to D'Arcy Wentworth Thompson. Clarendon Press, Oxford. 408 p.

Meek, A. 1916. The migration of fish. Arnold, London. 427. p.

Midttun, L. and O. Nakken. 1977. Some results of abundance estimation studies with echo integrators. Rapp. Procès-Verb. Cons. Intern. Explor. Mer, 170:253–58.

Midttun, L., and G. Saetersdal. 1957. On the use of echo-sounder observations for estimating fish abundance. Intern. N.W. Atlantic Fish., Spec. Pub. No. 2. 2 p.

Mitson, R. B., and T. J. Storeton-West. 1971. A transponding acoustic fish tag. Radio Electron. Eng., 41(11):483–89.

Moiseev, P. A. 1973. Development of fisheries for traditionally exploited species. J. Fish. Res. Board, Can., 30(12):2109–20.

Møller, D. 1968. Genetic diversity in spawning cod along the Norwegian coast. Hereditas, 60:1–32.

Møller, D. 1970. Transferrin polymorphism in Atlantic salmon (*Salmo salar*). J. Fish. Res. Board, Can., 27:1617–25.

Møller, D. 1971. Preliminary results of an Atlantic salmon population study. Rapp. Procès-Verb. Cons. Intern. Explor. Mer, 161:96.

Moran, P. A. P. 1962. The statistical processes of evolutionary theory. Oxford Univ. Press, London. 208 p.

Mortensen, E. 1977. Density dependent mortality of trout fry (*Salmo trutta* L.) and its relationship to the management of small streams. J. Fish. Biol., 11(6):613–17.

Motoda, S., and Y. Hirano. 1963. Review of Japanese herring investigations. Rapp. Procès-Verb. Cons. Intern. Explor. Mer, 154:249–62.

Murphy, G. I. 1965. A solution of the catch equation. J. Fish. Res. Board, Can., 22(1):191–202.

Murphy, G. I. 1966. Population biology of the Pacific sardine (*Sardinops caerulea*). Calif. Acad. Sci., Proc., Ser. 4, Vol. 34(1):1–84.

Murphy, G. I., and R. S. Shomura. 1955. Longline fishing for deep-swimming tunas in the central Pacific, August-November 1952. U.S. Fish and Wildlife Serv., Spec. Sci. Rep. No. 137. 42 p.

Muzinic, R., and B. B. Parish. 1960. Some observations on the body proportions of North Sea autumn spawning herring. J. Cons. Intern. Explor. Mer, 25(2):191–203.

Nakamura, H. 1969. Tuna distribution and migration. Fishing News (Books), London. 76 p.

Nakken, O., and K. Olsen. 1977. Target strength measurements of fish. Rapp. Procès-Verb. Cons. Intern. Explor. Mer, 170:52–69.

Neave, F. 1953. Principles affecting the size of pink and chum salmon populations in British Columbia. J. Fish. Res. Board, Can., 9:450–91.

Neess, R., and R. C. Dugdale. 1959. Computation of production for populations of aquatic midge larvae. Ecology, 40:425–30.

Nei, M. 1971. Interspecific gene differences and evolutionary time estimated from electrophoretic data on protein identity. Amer. Nat., 105:385–98.

Nei, M. 1972. Genetic distance between populations. Amer. Nat., 106:283–92.

Newman, G. G. 1970. Stock assessment of the Pilchard *Sardinops ocellata* at Walvis Bay, South West Africa. Invest. Rep. Div. Sea Fish. S. Afr., No. 85. 13 p.

Nikolskii, G. V. 1953. On some regularities of fecundity dynamics in fishes, p. 199–206. *In* his Papers on the general problems of ichthyology. Acad. Sci., U.S.S.R., 36.

Nikolskii, G. V. 1969. Theory of fish population dynamics. Oliver and Boyd, Edinburgh. 323 p.

O'Connell, C. P., and L. P. Raymond. 1970. The effect of food density on survival and growth of early post yolk-sac of the northern anchovy *Engraulis mordax* (Girard) in the laboratory. J. Exp. Mar. Biol. Ecol., 5(2):187–97.

Olson, F. C. S. 1964. The survival value of fish schooling. J. Cons. Intern. Explor. Mer, 29:115–16.

Otsu, T. 1960. Albacore migration and growth in the north Pacific Ocean as estimated from tag recoveries. Pacific Sci., 14(3):257–60.

Otsu, T., and R. N. Uchida. 1963. Model of the migration of albacore in the north Pacific Ocean. Fish. Bull. U.S. Fish and Wildlife Serv., 63(1):33–44.

Ottestad, P. 1969. Forecasting the annual yield in sea fisheries. Nature, 185(4707):183.

Outram, D. N. 1958. The magnitude of herring spawn losses due to bird predators on the west coast of Vancouver Island. Progr. Rep. Pac. Coast Stations, 3:9–13.

Paloheimo, J. E. 1961. Studies on estimation of mortalities. I. Comparison of a method described by Beverton and Holt and a new linear formula. J. Fish. Res. Board, Can., 18(5):645–62.

Paloheimo, J. E., and L. M. Dickie. 1964. Abundance and fishery success. Rapp. Procès-Verb. Cons. Intern. Explor. Mer, 155:152–63.

Paloheimo, Y., and L. M. Dickie. 1965. Food and growth of fishes. 1. A growth curve derived from experimental data. J. Fish. Res. Board, Can., 22:521–54.

Paloheimo, Y., and L. M. Dickie. 1966a. Food and growth of fishes. 2. Effects of food and temperature on the relation between metabolism and growth. J. Fish. Res. Board, Can., 23:869–908.

Paloheimo, Y., and L. M. Dickie. 1966b. Food and growth of fishes. 3. Relations among food, body size and growth efficiency. J. Fish. Res. Board, Can., 23:1209–48.

Parker, R. A. 1955. A method for removing the effect of recruitment on Petersen-type population estimates. J. Fish. Res. Board, Can., 12:447–50.

Parker, R. R., and P. A. Larkin. 1959. A concept of growth in fishes. J. Fish. Res. Board, Can., 16(5):721–45.

Parrish, B. B., and R. E. Craig. 1963. The herring of the northwestern North Sea — Post war changes in the stock fished by Scottish drifters. Cons. Intern. Explor. Mer, Herring Symp. [1961], 154:139–58.

Parrish, B. B., A. Saville, R. E. Craig, I. G. Baxter, and R. Priestley. 1959. Observations on herring spawning and larval distributions in the Firth of Clyde in 1958. J. Mar. Biol. Ass., U.K., N.S., 38:445–53.

Paulik, G. 1963. Estimates of mortality rates from tag recoveries. Biometrics, 19:28–57.

Pawson, M. G., S. T. Forbes, and J. Richards. 1975. Results of the 1975 acoustic surveys of blue whiting to the west of Britain. Intern. Council Explor. Sea, Council Mtg., H15. 5 p. Mimeograph.

Payne, R. H., A. R. Child, and A. Forrest. 1971. Geographical variation in the Atlantic salmon. Nature, 231:250–52.

Pearcy, W. G. 1962. Ecology of young winter flounder in an estuary. Peabody Mus. Natur. Hist., Yale Univ., New Haven, Conn., Bull. Bingham Oceanogr. Coll., 18(1):1–78.

Pearl, R. 1930. The biology of population growth. Knopf, New York. 330 p.

Pella, J. J., and P. K. N. Tomlinson. 1969. A generalized stock production model. Bull. Inter-Amer. Trop. Tuna Comm., 13:421–96.

Petersen, C. G. J. 1894. On the biology of our flatfishes and on the decrease of our flatfisheries. Rep. Dansk. Biol. Sta., 4. 146 p.

Petersen, C. G. J. 1896. The yearly immigration of young plaice into the Limfjord from the German Sea. Rep. Dansk. Biol. Sta. [1895], 6:1–48.

Pinhorn, A. T. 1975. Estimates of natural mortality for the cod stock complex in ICNAF Divisions 2J, 3K and 3L. Res. Bull. Intern. Comm. N.W. Atlantic Fish., 11:31–36.

Pitt, T. 1973. Assessment of American plaice stocks in ICNAF Divisions 3L and 3N. Res. Bull. Intern. Comm. N.W. Atlantic Fish., 10:63–77.

Platt, T. 1975. Analysis of the importance of spatial and temporal heterogeneity in the estimation of annual production by phytoplankton in a small enriched marine basin. J. Exp. Mar. Biol. Ecol., 18(2):99–109.

Pope, J. G. 1971. An investigation of the accuracy of virtual population analysis. Res. Doc. Intern. Comm. N.W. Atlantic Fish., 71/116. 11 p. Mimeograph.

Pope, J. G. 1972. An investigation of the accuracy of virtual population analysis using cohort analysis. Res. Bull. Intern. Comm. N.W. Atlantic Fish., 9:65–74.

Pope, J. G. 1976a. The application of mixed fisheries theory of the cod and redfish stocks of sub area Z and Division 3K. Selected papers, 1976, Intern. Comm. N.W. Atlantic Fish., 1:163–69.

Pope, J. G. 1976b. The effect of biological interaction on the theory of mixed fisheries. Selected papers, 1976, Intern. Comm. N.W. Atlantic Fish., 1:157–62.

Pope, J. G., and D. J. Garrod. 1975. Sources of error in catch and effort quota regulations with particular reference to variations in the catchability coefficient. Res. Bull. Intern. Comm. N.W. Atlantic Fish., 11:17–30.

Postuma, K. H. 1963. The catch per unit effort and mortality rates in the Southern Bight and Channel herring fisheries. Rapp. Procès-Verb. Cons. Intern. Explor. Mer, 154:190–97.

Poulsen, E. M. 1930a. On the fluctuations in the abundance of cod fry in the Kattegat and the Belt Sea and causes of the same. Rapp. Procès-Verb. Cons. Intern. Explor. Mer, 65:26–30.

Poulsen, E. M. 1930b. Investigations of fluctuations in the cod stock in Danish waters. Rapp. Procès-Verb. Cons. Intern. Explor. Mer, 68:20–23.

Pritchard, A. L. 1938. Transplantation of pink salmon (Oncorhynchus gorbuscha) into Masset inlet, British Columbia, in the barren years. J. Biol. Board, Can., 4:141–50.

Pritchard, A. L. 1939. Homing tendency and age at maturity of pink salmon (Oncorhynchus gorbuscha) in British Columbia. J. Fish. Res. Board, Can., 4:233–51.

Pritchard, A. L. 1948. A discussion of the mortality in pink salmon (Oncorhynchus gorbuscha) during their period of marine life. Roy. Soc. Can., Trans., Ser. 3, Vol. 42:125–33.

Probst, R. T., and E. L. Cooper. 1954. Age, growth and production of the lake sturgeon (Acipenser fulvescens) in the Lake Winnebago region, Wisconsin. Trans. Amer. Fish. Soc., 84:207–27.

Purdom, C. E., D. Thompson, and P. R. Dando. 1976. Genetic analysis of enzyme polymorphisms in plaice (Pleuronectes platessa). Heredity, 37(2):193–206.

Purdom, C. E., and T. Wyatt. 1969. Racial differences in Irish and North Sea plaice (Pleuronectes platessa). Nature, 222:780–88.

Quinn, W. H. 1974. Outlook for El Niño-type conditions in 1975. NORPAX Highlights, 2(6):2–3.

Raitt, D. S. 1939. The rate of mortality of the haddock of the North Sea stock, 1919–1938. Rapp. Procès-Verb. Cons. Intern. Explor. Mer, 110:65.

Raitt, D. F. S. 1968. The population dynamics of the Norway pout in the North Sea. Dept. Agric. Fish. Scotland, Mar. Res.,. 24 p.

Ramster, J. W., T. W. Wyatt, and R. G. Houghton. 1973. Towards a measure of the rate of drift of planktonic organisms in the vicinity of the Straits of Dover. Intern. Council Explor. Sea, Council Mtg., L8. 14 p. Mimeograph.

Rao, C. R. 1952. Advanced statistical methods in biometric research. John Wiley, New York. 390 p.

Reid, J. L. 1962. On circulation, phosphate phosphorous content, and zoopopulation volumes in the upper part of the Pacific Ocean. Limnol. Oceanogr., 7(3):287–306.

Richards, F. J. 1959. A flexible growth function for empirical use. J. Exp. Bot., 10:290–300.

Richardson, I. D., D. H. Cushing, F. R. Harden Jones, R. J. H. Beverton, and R. W. Blacker. 1959. Echo sounding experiments in the Barents Sea. H. M. Stationery Off., London, Fish. Invest., Ser. 2, Vol. 22(9). 55 p.

Ricker, W. E. 1940. Relation of "catch per unit effort" to abundance and rate of exploitation. J. Fish. Res. Board, Can., 5:43–70.

Ricker, W. E. 1945. Abundance, exploitation and mortality of the fishes in two lakes. Invest. Indiana Lakes and Streams, 2(17):345–448.

Ricker, W. E. 1946. Production and utilization of fish populations. Ecol. Monogr., 16:373–91.

Ricker, W. E. 1948. Methods of estimating vital statistics of fish populations. Bloomington, Indiana Univ. Pub., Sci. Ser., No. 15. 101 p.

Ricker, W. E. 1950. Cycle dominance among the Fraser sockeye. Ecology, 31(1):6–26.

Ricker, W. E. 1954. Stock and recruitment. J. Fish. Res. Board, Can., 11(5):559–623.

Ricker, W. E. 1958a. Handbook of computations for biological statistics of fish populations. Bull. Fish. Res. Board, Can., Bull. No. 119. 300 p.

Ricker, W. E. 1958b. Maximum sustained yields from fluctuating environments and mixed stocks. J. Fish. Res. Board, Can., 15(5):991–1006.

Ricker, W. E. 1973. Critical statistics from two reproduction curves. Rapp. Procès-Verb. Cons. Intern. Explor. Mer, 164:333–40.

Ricker, W. E. 1975. Computation and interpretation of biological statistics of fish populations. Bull. Fish. Mar., Ser 5, Canada, No. 191. 382 p.

Ricker, W. E., and R. E. Foerster. 1948. Computation of fish production. Peabody Mus. Natur. Hist., Yale Univ., New Haven, Conn., Bull. Bingham Oceanogr. Coll., 11(4):173–211.

Ridgway, G. J., R. D. Lewis, and S. W. Sherburne. 1971. Serological and biochemical studies tions of sockeye salmon, Oncorhynchus nerka. U.S. Fish and Wildlife Serv., Spec. Sci. Rep. — Fish., No. 257. 9 p.

Ridgway, G. J., R. D. Lewis, and S. W. Sherburne. 1971. Serological and biochemical studies of herring populations in the Gulf of Maine. Rapp. Procès-Verb. Cons. Intern. Explor. Mer, 161:21–25.

Riley, J. D., and Corlett, J. 1965. The numbers of 0-group plaice in Port Erin Bay, 1964–66, p. 51–56. In Ann. Rep. No. 78, Mar. Biol. Ass. Port Erin.

Robinson, B. J. 1978. In situ measurements of fish target strength. Proc. Conf. Acoustics in Fisheries, Institute of Acoustics. Univ. Bath. 7 p.

Robson, D. S. 1963. Maximum likelihood estimation of a sequence of annual survival rates from a capture-recapture series. Intern. Comm. N.W. Atlantic Fish., Spec. Pub. No. 4:330–35.

Robson, D. S., and H. A. Régier. 1964. Sample size in Petersen mark recapture experiments. Trans. Amer. Fish. Soc., 93:215–26.

Rodino, E., and A. Comparini. 1979. Genetic variability in the European eel Anguilla anguilla L. In B. Battaglia and J. Beardmore [eds.], Marine organisms: genetics, ecology and evolution. Plenum Press, New York.

Rogers, J. S. 1972. Measures of genetic similarity and genetic distance. Studies in genetics, VII, Univ. Texas Pub. 7213:145–53.

Rollefsen, G. 1934. The cod otolith as a guide to race, sexual development and mortality. Rapp. Procès-Verb. Cons. Intern. Explor. Mer, 88(2):1–5.

Rollefsen, G. 1953. The selectivity of different fishing gear used in Lofoten. J. Cons. Intern. Explor. Mer, 19(2):191–94.

Rollefsen, G. 1954. Observations on the cod and cod fisheries of Lofoten. Rapp. Procès-Verb. Cons. Intern. Explor. Mer, 136:40–47.

Rollefsen, G. 1955. The arctic cod. U.N. Sci. Conf. Conserv. Utiliz. Resources, Proc., Rome, p. 115–17.

Rosenthal, H., and G. Hempel. 1971. Experimental estimates of minimum food density for herring larvae. Rapp. Procès-Verb. Cons. Intern. Explor. Mer, 160:125–27. Oliver Boyd, Edinburgh. 552 p.

Rosenthal, H., G. Hempel. 1971. Experimental estimates of minimum food density for herring larvae. Rapp. Procès-Verb. Cons. Intern. Explor. Mer, 160:125–27.

Rothschild, B. J. 1967. Competition for gear in a multiple-species fishery. J. Cons. Intern. Explor. Mer, 31(1):102–10.

Rothschild, B. J. 1977. Fishing effort, p. 96–115. In J. A. Gulland [ed.], Fish population dynamics. John Wiley, New York. 372 p.

Rothschild, B. J., and J. W. Balsiger. 1971. A linear programming solution to salmon management. Fishery Bull., U.S. Dept. Comm., Nat. Mar. Fish. Serv., 69(1):117–40.

Rothschild, B. J., and M. Y. Y. Yong. 1970. Apparent abundance, distribution and migrations of albacore, *Thunnus alalunga*, in the North Pacific long line grounds. Spec. Sci. Rep., U.S. Fish. Wildl. Serv., Fisheries, 623:1–37.

Rounsefell, G. A. 1958. Factors causing decline in sockeye salmon of Karluk River, Alaska. U.S. Fish and Wildlife Serv., Fish. Bull. 58(130):83–169.

Runnstrøm, S. 1934. The Pelagic distribution of the herring larvae in the Norwegian waters. Rapp. Procès-Verb. Cons. Intern. Explor. Mer, 88(5), 6 p.

Runnstrøm, S. 1936. A study of the life history and migration of the Norwegian spring herring based on the analysis of the winter rings and summer zones of the scale. Fiskeridir. Skr. Ser. Havunders., 5(2). 103 p.

Runnstrøm, S. 1941. Quantitative investigations on herring spawning and its yearly fluctuations at the west coast of Norway. Fiskerid. Skr. Ser. Havunders., 6(8). 70 p.

Russell, E. S. 1931. Some theoretical considerations on the "overfishing" problem. J. Cons. Intern. Explor. Mer, 6(1):3–20.

Russell, F. S. 1973. A summary of the observations on the occurrence of planktonic stages of fish, Plymouth, 1924–1972. J. Mar. Biol. Ass., U.K., 53:346–55.

Saetersdal, G., and A. Hylen. 1959. Skreiundersøkelsene og skreifisket i 1959. Fisken og Havet., No. 1:1–18.

Sahrhage, D., and G. Wagner. 1978. On fluctuations in the haddock population of the North Sea. Rapp. Procès-Verb. Cons. Intern. Explor. Mer, 172:72–85.

Saila, S. B. 1961. A study of winter flounder movements. Limnol. Oceanogr., 6:292–98.

Saila, S. B., and J. M. Flowers. 1969. Elementary applications of search theory to fishing tactics as related to some aspects of fish behaviour. F.A.O. Fish. Rep., 62(2):343–56.

Saila, S. B., and R. A. Shappy. 1963. Random movements and orientation in salmon migration. J. Cons. Intern. Explor. Mer, 28:153–66.

Santander, H. 1976. La corriente Peruana. II. Aspectos biologicos. I.D.O.E. Workshop on El Niño, Guayaquil. F.A.O. Fish. Rep., 185:285–98.

Savage, R. E. 1937. The food of North Sea herring, 1930–1934. H. M. Stationery Off., London, Fish. Invest., Ser. 2, Vol. 15(5). 60 p.

Schaefer, M. B. 1951. Estimation of size of animal populations by marking experiments. Fish. Bull. U.S. Fish and Wildlife Serv., 52(69):191–203.

Schaefer, M. B. 1954. Some aspects of the dynamics of populations important to the management of the commercial fish populations. Bull. Inter-Amer. Trop. Tuna Comm., 1(2):27–56.

Schaefer, M. B. 1957. A study of the dynamics of the fishery for yellowfin tuna in the eastern tropical Pacific Ocean. Bull. Inter-Amer. Trop. Tuna Comm., 2:245–85.

Schaefer, M. B., and R. J. H. Beverton. 1963. Fishery dynamics — their analysis and interpretation, p. 464–83. *In* M. N. Hill [ed.], The sea. Vol. 2. Interscience Publishers, New York. 554 p.

Schmidt, J. 1909. The distribution of the pelagic fry and the spawning regions of the gadoids in the North Atlantic from Iceland to Spain. Rapp. Procès-Verb. Cons. Intern. Explor. Mer, Vol. 10B, Spec. part 4. 158 p.

Schmidt, J. 1914. First report on eel investigations 1913. Rapp. Procès-Verb. Cons. Intern. Explor. Mer, 18:1–30.

Schmidt, J. 1915. Second report on eel investigations, 1915. Rapp. Procès-Verb. Cons. Intern. Explor. Mer, 23:1–24.

Schmidt, J. 1917. Racial investigations. I. *Zoarces viviparus* L. and local races of the same. Comptes rendus Trav. Lab., Carlsberg, 13(3):279–396.

Schmidt, J. 1922. The breeding places of the eel. Roy. Soc. London, Phil. Trans., Ser. B, Vol. 211:179–208.

Schmidt, J. 1930. Racial investigations. X. The Atlantic cod (*Gadus callarias* L.) and local races of the same. Comptes rendus Trav. Lab., Carlsberg, 18(6):1–72.

Schumacher, A. 1971. Fishing mortality and stock size in the West Greenland cod. Res. Bull. Intern. Comm. N.W. Atlantic Fish., 8:15–20.

Scofield, W. L. 1929. Sardine fishing methods at Monterey, California. Calif. Dept. Nat. Res., Div. Fish and Game, Fish. Bull., No. 19. 61 p.

Seber, G. A. F. 1965. A note on the multiple recapture census. Biometrika, 52(1/2):249–59.

Sette, O. F. 1943. Biology of the Atlantic mackerel (*Scomber scombrus*) of North America. I. Early life history including the growth, drift and mortality of egg and larval populations. Fish. Bull. U.S. Fish and Wildlife Serv., 50(38):149–234.

Shapovalov, L. 1937. Trout and salmon marking in California. Calif. Fish and Game, 23:205–7.

Shepard, M. P., and F. C. Withler. 1958. Spawning stock size and resultant production for Skeena sockeye. J. Fish. Res. Board, Can., 15(5):1007–25.

Shepherd, J. G., and D. H. Cushing. In press. A mechanism for density dependent survival of larval fish as the basis of a stock-recruitment relationship. J. Cons. Intern. Explor. Mer.

Sick, K. 1962. Nye metoder fil raceundersogelser af fisk. Skr. Danm. Fisk. og Havunders., 22:13–16.

Sick, K., E. Bahn, O. Frydenberg, J. T. Nielsen, and D. von Wettstein. 1967. Haemoglobin polymorphism of the American fresh water eel *Anguilla*. Nature, 214(5093):1141–42.

Silliman, R. P., and F. N. Clark. 1945. Catch per-unit-of-effort in California waters of the sardine (*Sardinops caerulea*) 1932–1942. Calif. Dept. Nat. Res., Div. Fish and Game, Fish. Bull., No. 62. 78 p.

Simpson, A. C. 1959. The spawning of the plaice in the North Sea. H. M. Stationery Off., London, Fish. Invest., Ser. 2, Vol. 22(7). 100 p.

Sindermann, C. J. 1961. Serological techniques in fishery research. 26th N. Amer. Wildlife and Nat. Resources Conf., Trans., p. 298–309.

Sindermann, C. J. 1970. Principal diseases of marine fish and shellfish. Academic Press, New York. 369 p.

Sindermann, C. J., and D. F. Mairs. 1959. The C blood group system of Atlantic sea herring. Anat. Rec., 134:640.

Skud, B. E. 1975. Revised estimates of halibut abundance and the Thompson-Burkenroad debate. Sci. Rep. Intern. Pacif. Halibut Comm., 56. 36 p.

Slobodkin, L. B. 1966. Growth and regulation of animal populations. Holt, Rinehart and Winston, New York. 184 p.

Solomon, M. E. 1949. The natural control of animal populations. J. Anim. Ecol., 18:1–35.

Southward, G. M. 1967. Growth of Pacific halibut. Rep. Intern. Pacif. Halibut Comm., 43. 40 p.

Sprague, L. M., and A. M. Vrooman. 1962. A racial analysis of the Pacific sardine (*Sardinops caerulea*) based on studies of erythrocyte antigens. Ann. N.Y. Acad. Sci., 97:131–38.

Sproston, N. G. 1947. *Ichthyosporidium hoferi* (Plehn and Malsow, 1911) and internal fungoid parasite of the mackerel. J. Mar. Biol. Ass., U.K., N.S., 26:72–98.

Steele, J. H. 1961. The environment of a herring fishery. Fisheries, Scotland, Mar. Res. 1961, No. 6. 18 p.

Steele, J. H. 1974. The structure of marine ecosystems. Blackwell Scientific Publications, Oxford, 128 p.

Steele, J. H., and R. R. C. Edwards. 1969. The ecology of 0-group plaice and common dabs in Loch Ewe. IV. Dynamics of the plaice and dab populations. J. Exp. Mar. Biol. Ecol., 4:174–88.

Stommel, H. 1958. The gulf stream: A physical and dynamical description. Cambridge Univ. Press, London. 248 p.

Strasburg, D. W. 1959. An instance of natural mass mortality of larval frigate mackerel in the Hawaiian Islands. J. Cons. Intern. Explor. Mer, 24(2):255–63.

Stubbs, A. R., and R. G. G. Lawrie. 1962. Asdic as an aid to spawning ground investigations. J. Cons. Intern. Explor. Mer, 27(3):248–60.

Suda, A. 1963. Structure of the albacore stock and fluctuation in the catch in the North Pacific area. F.A.O. Fish. Rep. No. 6(3):1237–77.

Sund, O. 1935. Echo sounding in fishery research. Nature, 135(3423):953.

Sundnes, G. 1965. Energy metabolism and migration of fish. Intern. Comm. N.W. Atlantic Fish., Spec. Pub. No. 6:743–46.

Swallow, J. C., and L. V. Worthington. 1961. An observation of a deep counter current in the western North Atlantic. Deep Sea Res., 8:1–19.

Swingle, H. S., and E. V. Smith. 1943. The management of ponds with stunted fish populations. Trans. Amer. Fish. Soc., 71st Ann. Meeting, 102–5.

Taft, A. C., and L. Shapovalov. 1938. Homing instinct and straying among steelhead trout (Salmo gairdneri) and silver salmon (Oncorhynchus kisutch). Calif. Fish and Game, 24:118–25.

Tait, J. B. 1930. The surface drift in the northern and middle areas of the North Sea in the Faroe-Shetland Channel. II. Sect. 1. A cartographical analysis of the results of Scottish surface drift bottle experiments commenced in the year 1910. Fisheries, Scotland, Sci. Invest. 1930, No. 4. 56 p.

Tait, J. B. 1937. The surface water drift in the northern and middle areas of the North Sea and in the Faroe-Shetland Channel. II. Sect. 3. A cartographical analysis of the results of Scottish surface drift bottle experiments of the year 1912; with a discussion on some hydrographical and biological implications of the drift bottle results of 1910, 1911 and 1912, including a statement of a theory of the upper water circulation of the northern and middle North Sea. Fisheries, Scotland, Sci. Invest. 1937, No. 1. 60 p.

Tanaka, S. 1960. Studies on the dynamics and management of fish populations. Bull. Tokai Reg. Fish. Res. Lab., 28:1–200.

Tanaka, S. 1962. On the salmon stocks of the Pacific coast of the United States and Canada (views of the Japanese National Section on the abstention cases of the United States and Canada). Intern. N. Pacific Fish. Comm., Bull. No. 9:69–85.

Tanaka, S. 1974. Significance of egg and larval surveys in the studies of population dynamics of fish, p. 151–57. In J. H. S. Blaxter [ed.], The early life history of fish. Springer-Verlag, Berlin. 765 p.

Tåning, A. V. 1937. Some features in the migration of cod. J. Cons. Intern. Explor. Mer, 12:1–35.

Tåning, A. V., H. Einarsson, and J. Eggvin. 1955. Records from the month of June of the Norwegian-Icelandic herring stock in the open ocean. Ann. Biol., Cons. Intern. Explor. Mer, 12:165–67.

Taylor, C. C. 1962. Growth equations with metabolic parameters. J. Cons. Intern. Explor. Mer, 27:270–86.

Templeman, W. 1965. Relation of periods of successful year classes of haddock on the Grand Bank to periods of success of year classes for cod, haddock and herring in areas to the north and east. Intern. Comm. N.W. Atlantic Fish., Spec. Pub. No. 6:523–33.

Templeman, W. 1972. Year class success in some North Atlantic stocks of cod and haddock. Intern. Comm. N.W. Atlantic Fish., Spec. Pub. No. 8:223–39.

Templeman, W., and G. L. Andrews. 1956. Jellied condition in the American plaice Hippoglossoides platessoides (Fabricius). J. Fish. Res. Board, Can., 13(2):147–82.

Templeman, W., and V. M. Hodder. 1965. Distribution of haddock on the Grand Bank in relation to season, depth and temperature. Intern. Comm. N.W. Atlantic Fish., Spec. Pub. No. 6:171–88.

Templeman, W., and A. W. May. 1965. Research vessel catches of cod in Hamilton Inlet Bank area in relation to depth and temperature. Intern. Comm. N.W. Atlantic Fish., Spec. Pub. No. 6:149–66.

Tesch, F. W. 1977. The eel. Chapman and Hall, London. 434 p.

Thompson, D'Arcy W. 1917. On growth and form. Cambridge Univ. Press, London. 794 p.

Thompson, W. F. 1945. Effect of the obstruction at Hell's Gate on the sockeye salmon of the Fraser River. Bull. Intern. Pacif. Salmon Comm., No. 1. 175 p.

Thompson, W. F., and F. H. Bell. 1934. Biological statistics of the Pacific halibut fishery. 2. Effect of changes in intensity upon total yield and yield per unit of gear. Rep. Intern. Fish. Comm., No. 8. 49 p.

Thompson, W. F., and W. C. Herrington. 1930. Life history of the Pacific halibut. Rep. Intern. Fish. Comm., No. 2. 137 p.

Thorne, R. F. 1977. Acoustic assessment of Pacific hake and herring stocks in Puget Sound, Washington and Southeastern Alaska. Rapp. Procès-Verb. Cons. Intern. Explor. Mer, 170:265–78.

Tiews, K. 1963. Synopsis of biological data on bluefin tuna *Thunnus thynnus* (Linnaeus) 1758 (Atlantic and Mediterranean). F.A.O. Fish. Rep. No. 6. F.A.O. World Sci. Meet. Biol. Tunas and Related Species, Proc., Species Synopsis, 12:422–81.

Townsend, C. H. 1935. The distribution of certain whales as shown by logbook records of American whaleships. Zoologica, 19:1–50.

Trout, G. C. 1957. The Bear Island cod: Migration and movements. H. M. Stationery Off., London, Fish. Invest., Ser. 2, Vol. 21(6). 51 p.

Uda, M. 1952. On the relation between the variation of the important fisheries conditions and the oceanographical conditions in the adjacent waters of Japan. J. Tokyo Univ. Fish., 38(3):363–89.

Uda, M. 1959. Oceanographic seminars. 2. Watermass boundaries — ''Siome'': Frontal theory in oceanography. Fish. Res. Board, Can., MS Rep. Ser. (Oceanogr. and Limnol.), 51:10–20.

Ulltang, Ø. 1976. Catch per unit of effort in the Norwegian purse seine fishery for Atlanto-Scandian (Norwegian spring spawning) herring. F.A.O. Fish. Tech. Paper, No. 155:91–101.

Ursin, E. 1973. On the prey size preferences of cod and dab. Meddr. Danm. Fisk. og Havunders., N.S., 7:85–98.

Utter, F., W. Ames, and H. O. Hodgins. 1970. Transferrin polymorphism in Coho salmon (*Oncorhynchus kisutch*). J. Fish. Res. Board, Can., 27:2371–73.

Van Campen, W. G. (transl.) 1960. Japanese summer fishery for albacore (*Germo alalunga*). U.S. Fish and Wildlife Serv., Res. Rep., No. 52. 29 p.

Veen, J. F. de. 1962. On the subpopulations of plaice in the southern North Sea. Intern. Council Explor. Sea, Council Mtg., 94. 6 p. Mimeograph.

Veen, J. F. de. 1976. On changes in some biological parameters in the North Sea sole (*Solea solea* L.). J. Cons. Intern. Explor. Mer, 37:60–90.

Veen, J. F. de, and L. K. Boerema. 1959. Distinguishing southern North Sea populations of plaice by means of otolith characteristics. Cons. Intern. Explor. Mer, Near Northern Seas Comm., Paper No. 91. 5 p. Mimeograph.

Verhulst, P. F. 1838. Note sur la loi que la population suit dans son accroissement. Corresp. Math. Phys., 10:113–21.

Vildoso, A. 1976. Distribution de la fauna. I. Aspectos Biologicos. I.D.O.E. Workshop on El Niño, Guayaquil. F.A.O. Fish. Rep., 185:62–79.

Walford, L. A. 1946. A new graphic method of describing the growth of animals. Biol. Bull., 90(2):141–47.

Walsh, J. J. 1975. A spatial simulation model of the Peru upwelling ecosystem. Deep Sea Res., 22(4):201–36.

Walters, C. J., and R. Hilborn. 1975. Adaptive control of fishing systems. Intern. Inst. Applied Systems Analysis, Res. Rep. 75–39. 38 p.

Ward, F. J., and P. A. Larkin. 1964. Cyclic dominance in Adams River sockeye salmon. Prog. Rep. 11, Intern. Pacif. Salmon Comm. 116 p.

Ware, D. M. 1975. Relation between egg size, growth and natural mortality of larval fish. J. Fish. Res. Board, Can., 32:2503–12.

Warren, C. E., and G. E. Davis. 1967. Laboratory studies on the feeding, bioenergetics and growth of fish, p. 175–214. In S. D. Gerking [ed.], Biological basis of freshwater fish production. Oxford Univ. Press, London.

Webb, P. W. 1975. Hydrodynamics and energetics of fish propulsion. Bull. Fish. Res. Board, Can., No. 190. 158 p.

Weihs, D. 1973. Optimal fish cruising speed. Nature, 245:48–50.

Weihs, D. 1978. Tidal transport as an efficient method for migration. J. Cons. Intern. Explor. Mer, 38:106–13.

Wimpenny, R. S. 1953. The plaice, being the Buckland lectures for 1949. Arnold, London. 144 p.

Winberg, G. G. 1956. Rate of metabolism and food requirements of fishes. Nauch. Trud. Belorusskovo Gos. Univ. i meni. V. I. Lenina, Minsk. 253 p.

Winters, G. H., and V. M. Hodder. 1975. Analysis of the Southern Gulf of St. Lawrence herring stock and implications concerning its future management. Res. Bull. Intern. Comm. N.W. Atlantic Fish., 11:43–60.

Wood, H. 1937. Movements of herring in the northern North Sea. Fisheries, Scotland, Sci. Invest. 1937, No. 3. 49 p.

Woodhead, P. M. J., and A. D. Woodhead. 1959. The effects of low temperatures on the physiology and distribution of the cod, Gadus morhua L., in the Barents Sea. Zool. Soc. London, Proc., 133(2):181–99.

Wooster, W., and O. Guillen. 1974. Characteristics of El Niño in 1972. J. Mar. Res., 32:387–404.

Wyatt, T. 1972. Some effects of food density on the growth and behaviour of plaice larvae. Mar. Biol., 14(3):210–16.

Wyrtki, K. 1966. Oceanography of the eastern equatorial Pacific Ocean. Oceanogr. Mar. Biol. Ann. Rev., 4:33–68.

Wyrtki, K. 1973. Teleconnections in the Equatorial Pacific Ocean. Science, 180:66–68.

Yamanaka, I. 1960. Comparative study of the population size of Japanese and California sardine. F.A.O. World Sci. Meet. Biol. Sardines and Related Species, Proc., 3:1151–92.

Zenkman, V. S., and A. F. Lysenko. 1978. Genetics and morphological differentiation of the North Sea immature herring. In Regularities in feeding and formation of the Atlantic commercial concentrations. Trudy Atlant. nauchno-issled. Inst. morsk. ryb. khoz. Okeanogr., 74:3–17.

Zijlstra, J. J. 1958. On the herring "races" spawning in the southern North Sea and English Channel (preliminary report). Rapp. Procès-Verb. Cons. Intern. Explor. Mer, 143(2):134–45.

Zijlstra, J. J. 1972. On the importance of the Wadden Sea as a nursery area in relation to the conservation of the southern North Sea fishery resources. Symp. Zool. Soc. Lond., 29:233–58.

Zupanovitch, S. 1968. Causes of fluctuations in sardine catches along the eastern coast of the Adriatic Sea. Anali Jadranskog Instituta, 4:401–89.

Zweifel, J. R. and R. Lasker. 1976. Prehatch and posthatch growth of fishes — a general model. Fish. Bull. U.S. Dept. Commerce, 74:609–21.

Acknowledgments

Figure 1. M. Graham, 1958. Fish population assessment by inspection, p. 67–68. *In* Some problems for biological fishery survey and techniques for their solution. Intern. Comm. N.W. Atlantic Fish., Spec. Pub. No. 1.

Figures 3 (left and right), 59, 79 (top), 80, 81. R. J. H. Beverton and S. J. Holt, 1957. On the dynamics of exploited fish populations. Fish. Invest., Ser. 2, Vol. 19, 533 p. With the permission of the Controller of Her Britannic Majesty's Stationery Office, London.

Figure 4. D. H. Cushing, 1957. The number of pilchards in the Channel. Fish. Invest., Ser. 2, Vol. 21(5), 27 p. With the permission of the Controller of Her Britannic Majesty's Stationery Office, London.

Figure 5. A. C. Hardy, G. T. D. Henderson, C. E. Lucas, and J. H. Fraser, 1936. The ecological relations between the herring and the plankton investigated with the plankton indicator. J. Mar. Biol. Ass. U.K., N.S., 21:147–304. With the permission of the Cambridge University Press.

Figures 6, 62. J. Ancellin and C. Nédelèc, 1959. Marquage de harengs en Mer du Nord et en Manche orientale (Campagne du "Président Théodore Tissier," Novembre 1957). Rev. Trav. Inst. Pêches Marit., 23:177–201.

Figure 8. W. C. Hodgson, 1957. The herring and its fishery. Routledge and Kegan Paul, Ltd., London. 197 p.

Figures 10, 13, 14. F. R. Harden Jones, 1968. Fish migration. Arnold, London. 325 p.

Figure 11. J. C. Corlett, 1958. Distribution of larval cod in the western Barents Sea, p. 281–88. *In* Some problems for biological fishery survey and techniques for their solution. Intern. Comm. N. W. Atlantic Fish., Spec. Pub. No. 1.

Figure 12. J. Schmidt, 1922. The breeding places of the eel. Roy. Soc. London, Phil. Trans., Ser. B, Vol. 211:179–208.

Figures 15, 54. A. C. Simpson, 1959. The spawning of the plaice in the North Sea. Fish. Invest., Ser. 2, Vol. 22(7), 100 p. With the permission of the Controller of Her Britannic Majesty's Stationery Office, London.

Figure 16. D. H. Cushing, 1972a. The production cycle and numbers of mature fish. Symp. Zool. Soc., London, 29:213–32. *Also* D. H. Cushing, 1975b. Marine ecology and fisheries. Cambridge University Press, Cambridge. 277 p.

Figure 17. J. F. de Veen and L. K. Boerema, 1959. Distinguishing southern North Sea populations of plaice by means of otolith characteristics. Cons. Intern. Explor. Mer, Near Northern Seas Comm., Paper No. 91. 5 p. Mimeograph.

Figures 20, 79 (bottom). D. H. Cushing and J. P. Bridger, 1966. The stock of herring in the North Sea and changes due to fishing. Fish. Invest., Ser. 2, Vol. 25(1), 123 p. With the permission of the Controller of Her Britannic Majesty's Stationery Office, London.

Figure 21. F. R. Harden Jones, M. Greer Walker, and G. P. Arnold, 1978. Tactics of fish movement in relation to migration strategy and water circulation, p. 185–207. *In* H. Charnock and Sir George Deacon [eds.], Advances in Oceanography. Plenum Press, New York.

Figure 22. W. F. Thompson and W. C. Herrington, 1930. Life history of the Pacific halibut. Rep. Intern. Fish. Comm. No. 2. 137 p.

Figure 23. D. H. Cushing, 1975b. Marine ecology and fisheries. Cambridge University Press, Cambridge. 277 p.

Figure 24. G. Saetersdal and A. Hylen, 1959. Skreiundersøkelsene og skreifisket i 1959. Fisken og Havet., No. 1:1–18.

Figure 25. O. Bostrøm, 1955. "Peder Ronnestad" Ekkolodding og meldetjeneste av Skreiforekomstene i Lofoten i tiden 1 March–2 Apr. 1955: Praktiske fiskeforsøk 1954 og 1955. Arsberet. Vedkomm. Norges Fisk., 9:66–70.

Figure 26. J. Hjort, 1914. Fluctuations in the great fisheries of northern Europe viewed in the light of biological research. Rapp. Procès–Verb. Cons. Intern. Explor. Mer, 20:1–228.

Figure 27. G. Dannevig, 1954. The feeding grounds of the Lofoten cod. Rapp. Procès–Verb. Cons. Intern. Explor. Mer, 136:87–102.

Figure 28. I. D. Richardson, D. H. Cushing, F. R. Harden Jones, R. J. H. Beverton, and R. W. Blacker, 1959. Echo sounding experiments in the Barents Sea. Fish. Invest., Ser. 2, Vol. 22(9), 55 p. With the permission of the Controller of Her Britannic Majesty's Stationery Office, London.

Figure 29. A. Hylen, L. Midttun, and G. Saetersdal, 1961. Torskeundersøkelsene i Lofoten og i Barentshavet 1960. Fisken og Havet., No. 2:1–14.

Figures 30, 47. A. C. Burd and D. H. Cushing, 1962. I. Growth and recruitment in the herring of the southern North Sea. II. Recruitment to the North Sea herring stocks. Fish. Invest. Ser. 2, Vol. 23(5), 71 p. With the permission of the Controller of Her Britannic Majesty's Stationery Office, London.

Figure 31. T. Otsu, 1960. Albacore migration and growth in the North Pacific Ocean as estimated from tag recoveries. Pacific Sci. 14(3):257–60.

Figure 32. J. F. de Veen, 1962. On the subpopulations of plaice in the southern North Sea. Intern. Council Explor. Sea, Council Mtg., 94. 6 p. Mimeograph.

Figure 33. W. de Ligny, 1969. Serological and biochemical studies on fish populations. Oceanogr. Mar. Biol., 7:411–513. *Also* D. H. Cushing, 1975b. Marine ecology and fisheries. Cambridge University Press, Cambridge. 277 p.

Figure 34. A. Jamieson and R. J. Turner, 1979. The extended series of Tf alleles in Atlantic cod (*Gadus morhua* L.), p. 699–729. *In* B. Battaglia and J. Beardmore [eds.], Marine organisms: Genetics, ecology, and evolution. Plenum Press, New York.

Figure 35. D. H. Cushing, 1975b. Marine ecology and fisheries. Cambridge University Press, Cambridge. 277 p.

Figures 36, 37. R. J. H. Beverton and S. J. Holt, 1959. A review of the lifespans and mortality rates of fish in nature and the relation to growth and other physiological characteristics, p. 142–77. *In* Ciba Foundation, Colloquia on ageing. V. The lifespan of animals. J. & A. Churchill Ltd., London.

Figures 38, 40, 41. V. S. Ivlev, 1961. The experimental ecology of the feeding of fishes. Yale Univ. Press, New Haven. 302 p.

Figure 39. D. H. Cushing, 1964. The work of grazing in the sea, p. 207–25. *In* D. J. Crisp [ed.], Grazing in terrestrial and marine environments. Blackwell, London.

Figure 44. R. J. H. Beverton and S. J. Holt, 1957. On the dynamics of exploited fish populations. H. M. Stationery Office, London, Fish. Invest., Ser. 2, Vol. 19. 533 p. R. C. A. Bannister, personal communication.

Figure 45. G. M. Southward, 1967. Growth of Pacific halibut. Rep. Intern. Pacif. Halibut Comm., 43. 40 p.

Figure 46. D. H. Cushing, 1960. The East Anglian fishery in 1959. World Fishing, 9(8):51–58.

Figure 49. D. M. Ware, 1975. Relation between egg size, growth and natural mortality of larval fish. J. Fish. Res. Board, Can., 32:2503–12.

Figure 50. E. Mortensen, 1977. Density dependent mortality of trout fry (*Salmo trutta* L.) and its relationship to the management of small streams. J. Fish. Biol. 11(6):613–17.

Figure 52. G. Rollefsen, 1953. The selectivity of different fishing gear used in Lofoten. J. Cons. Intern. Explor. Mer, 19(2):191–94.

Figure 55. Ø. Ulltang, 1976. Catch per unit of effort in the Norwegian purse seine fishery for Atlanto-Scandian (Norwegian spring spawning) herring. F.A.O. Fish. Tech. Paper, No. 155:91–101.

Figure 56. S. Tanaka, 1962. On the salmon stocks of the Pacific coast of the United States and Canada (views of the Japanese National Section on the abstention cases of the United States and Canada). Intern. N. Pacific Fish. Comm., Bull. No. 9:69–85.

Figure 57. O. Nakken and K. Olsen, 1977. Target strength measurements of fish. Rapp. Procès-Verb. Cons. Intern. Explor. Mer, 170:52–69.

Figure 58. D. H. Cushing, 1968b. Direct estimation of a fish population acoustically. J. Fish. Res. Board, Can. 25(11):2349–64. *Also* D. H. Cushing, 1975b. Marine ecology and fisheries. Cambridge University Press, Cambridge. 277 p.

Figure 60. D. J. Garrod, 1963. An estimation of the mortality rates in a population of *Tilapia esculenta* Graham (Pisces, Cichlidae) in Lake Victoria, East Africa. J. Fish. Res. Board, Can., 20:195–227.

Figure 61. J. A. Gulland, 1963. The estimation of fishing mortality from tagging experiments. Intern. Comm. N.W. Atlantic Fish., N. Atlantic Fish Marking Symp., Spec. Pub. No. 4:218–27.

Figure 63. L. M. Dickie, 1963. Estimation of mortality rates of Gulf of St. Lawrence cod from results of a tagging experiment. Intern. Comm. N.W. Atlantic Fish., N. Atlantic Fish Marking Symp., Spec. Pub. No. 4:71–80.

Figure 64. D. H. Cushing, 1975a. The natural mortality of the plaice. J. Cons. Intern. Explor. Mer, 36(2):150–157. *Also* D. H. Cushing, 1975b. Marine ecology and fisheries. Cambridge University Press, Cambridge. 277 p.

Figure 65a. I. Yamanaka, 1960. Comparative study of the population size of Japanese and California sardine. F.A.O. World Sci. Meet. Biol. Sardines and Related Species. Proc., 3:1151–92.

Figure 65b. S. Motoda and Y. Hirano, 1963. Review of Japanese herring investigations. Rapp. Procès–Verb. Cons. Intern. Explor. Mer, 154:249–62.

Figure 65c. R. Jordan, 1976. Biologia de la anchoveta. I. Resumen del conocimiento actual. I.D.O.E. Workshop on El Niño, Guayaquil. F.A.O. Fish. Rep., 185:359–99.

Figures 66, 68. R. J. H. Beverton, 1962. Long–term dynamics of certain North Sea fish populations, p. 242–64. *In* E. D. Le Cren and M. W. Holdgate [eds.], The exploitation of natural animal populations. Blackwell, London. 399 p.

Figure 67. Courtesy of D. J. Garrod.

Figure 69. R. C. A. Bannister, D. Harding, and S. J. Lockwood, 1974. Larval mortality and subsequent year–class strength in the plaice (*Pleuronectes platessa* L.), p. 21–37. *In* J. H. S. Blaxter [ed.], The early life history of fish. Springer–Verlag, Berlin.

Figures 71, 72. J. H. Steele and R. R. C. Edwards, 1969. The ecology of O-group plaice and common dabs in Loch Ewe. IV. Dynamics of the plaice and dab populations. J. Exp. Mar. Biol. Ecol., 4:174–88.

Figure 73. W. E. Ricker, 1958a. Handbook of computations for biological statistics of fish populations. Fish. Res. Board, Can., Bull. No. 119. 300 p.

Figure 74a. D. H. Cushing and J. G. K. Harris, 1973. Stock and recruitment and the problem of density dependence. Rapp. Procès–Verb. Cons. Intern. Explor. Mer, 164:142–55.

Figure 74b, c, e, f, g. D. H. Cushing and J. G. K. Harris, 1973. Stock and recruitment and the problem of density dependence. Rapp. Procès–Verb. Cons. Intern. Explor. Mer, 164:142–55.

Figure 74d. S. Tanaka, 1974. Significance of egg and larval surveys in the studies of population dynamics of fish, p. 151–57. *In* J. H. S. Blaxter [ed.], The early life history of fish. Springer–Verlag, Berlin. 765 p.

Figure 74h. D. Sahrhage and G. Wagner, 1978. On fluctuations in the haddock population of the North Sea. Rapp. Procès–Verb. Cons. Intern. Explor. Mer, 172:72–85.

Figure 75a. R. S. Wimpenny, 1953. The plaice, being the Buckland lectures for 1949. Arnold, London, 144 p.

Figure 75b. Y. Fukuda, 1962. On the stocks of halibut and their fisheries in the Northeast Pacific. Intern. N. Pacific Fish. Comm., Bull No. 7:39–50.

Figure 75c. D. H. Cushing and J. P. Bridger, 1966. The stock of herring in the North Sea and changes due to fishing. H. M. Stationery Off., London, Fish. Invest., Ser. 2, Vol. 25(1). 123 p.

Figure 75d. International Commission on Whaling, 1964. Fourteenth Report of the Commission. Intern. Whaling Comm., London. 122 p.

Figure 75a–d. *Also* D. H. Cushing, 1975b. Marine ecology and fisheries. Cambridge University Press, Cambridge. 277 p.

Figure 76. M. B. Schaefer, 1957. A study of the dynamics of the fishery for yellowfin tuna in the eastern tropical Pacific Ocean. Bull. Inter–Amer. Trop. Tuna Comm. 2:245–85. *Also* D. H. Cushing, 1975b. Marine ecology and fisheries. Cambridge University Press, Cambridge. 277 p.

Figure 77. C. J. Walters and R. Hilborn, 1975. Adaptive control of fishing systems. Intern. Inst. Applied Systems Analysis, Res. Rep. 75–39. 38 p.

Figure 78. J. A. Gulland, 1961. Fishing and the stocks of fish at Iceland. Fish. Invest., Ser. 2, Vol. 23(4), 52 p. With the permission of the Controller of Her Britannic Majesty's Stationery Office, London.

Figure 82. D. H. Cushing, 1959a. On the effect of fishing on the herring of the Southern North Sea. J. Cons. Intern. Explor. Mer, 24(2):283–307.

Figure 83. A. D. Clayden, 1972. Simulation of the changes in abundance of the cod (*Gadus morhua* L.) and the distribution of fishing in the North Atlantic. Fish. Invest., Lond., Ser. 2, Vol. 27(1). 58 p. *Also* D. H. Cushing, 1975b. Marine ecology and fisheries. Cambridge University Press, Cambridge. 277 p.

Figure 84b. D. H. Cushing, 1980. The decline of the herring stocks and the gadoid outburst. J. Cons. Intern. Explor. Mer, 39(1):74–85.

Figure 85. K. M. Brander, 1977. The management of Irish Sea fisheries: A review. Min. Agr., Fish. Food, London, Lab. Leaflet No. 36. 40 p.

Figures 86, 87. A. J. Lee, 1952. The influence of hydrography on the Bear Island cod fishery. Rapp. Procès–Verb. Cons. Intern. Explor. Mer, 131:74–102.

Figure 88. A. V. Tåning, H. Einarrsson, and J. Eggvin, 1955. Records from the month of June of the Norwegian-Icelandic herring stock in the open ocean. Ann. Biol., Cons. Intern. Explor. Mer, 12:165–67.

Figure 89. J. E. King and T. S. Hida, 1957. Zooplankton abundance in the central Pacific. II. U.S. Fish and Wildlife Serv., Fish. Bull., 57(118):365–95.

Figure 90. C. H. Townsend, 1935. The distribution of certain whales as shown by logbook records of American whaleships. Zoologica, 19:1–50.

Figure 91. T. J. Hart and R. I. Currie, 1960. The Benguela current. Discovery Reports, Vol. 31:1–297. With the permission of the National Institute of Oceanography (now the Institute of Oceanographic Sciences, U.K.).

Figure 92. K. Wyrtki, 1966. Oceanography of the eastern equatorial Pacific Ocean. Oceanogr. Mar. Biol. Ann. Rev., 4:33–68.

Figure 93. J. L. Reid, 1962. On circulation, phosphate phosphorus content, and zooplankton volumes in the upper part of the Pacific Ocean. Limnol. and Oceanogr., 7(3):287–306.

Figure 94. D. H. Davies, 1958. The South African pilchard (*Sardinops ocellata*): The predation of sea birds in the commercial fishery. Dep. Comm. Ind., Div. Fish., Union South Africa, Invest. Rep., No. 31:1–15.

Figure 95. P. Ottestad, 1969. Forecasting the annual yield in sea fisheries. Nature, 185 (4707)183.

Figures 96, 97, 100. D. H. Cushing and R. R. Dickson, 1976. The biological response in the sea to climatic changes. Adv. Mar. Biol., 14:1–122.

Figure 98. F. S. Russell, 1973. A summary of the observations on the occurrence of planktonic stages of fish off Plymouth, 1924–72. J. Mar. Biol. Ass. U.K., NS 53:346–355.

Figure 99. D. H. Cushing, 1969. The regularity of the spawning season of some fishes. J. Cons. Intern. Explor. Mer, 33(1):81–92.

Figure 101. F. R. Harden Jones, G. P. Arnold, M. Greer Walker, and P. Scholes, 1979. Selective tidal stream transport and the migration of the plaice (*Pleuronectes platessa* L.) in the Southern North Sea. J. Cons. Intern. Explor. Mer, 38(1):331–37.

Figure 102. V. S. Zenkman and A. F. Lysenko, 1978. Genetics and morphological differentiation of the North Sea immature herring. *In* Regularities in feeding and formation of the Atlantic commercial concentrations. Trudy Atlant. nauchno-issled. Inst. morsk. ryb. khoz. Okeanogr., 74:3–17.

Figure 103. J. G. Shepherd and D. H. Cushing (in press). A mechanism for density dependent survival of larval fish as the basis of a stock–recruitment relationship. J. Cons. Intern. Explor. Mer.

Index

Abramis brama Linnaeus, 84

Absorption: efficiency, 74; of water by eggs, 151

Accessibility of fish to capture: vertically, in sprats, 104; horizontally, in albacore, 104

Acipenser fulvescens Rafinesque, 77

Acipenser stellatus Pall, 77

Acoustic methods: to estimate stock, 112, 239; first echo record of cod in Vest Fjord, 113; hake in Port Susan, Puget Sound, 113; for single fishes, dependence of target strength on length, 113–14

Acoustic survey: of hake off South Africa and Namibia, 114, 115; Pacific hake, southeast Alaskan herring, sockeye salmon, juvenile fish in Barents Sea, anchovy and horse mackerel, blue whiting, 117

Adams River stock of sockeye salmon, 161

Admiralty Research Laboratory (ARL) scanner: and plaice migration, 34, 233; used to estimate trawl efficiency for plaice, 112

Adriatic sardine, 225

Age distribution: in East Anglian herring fishery, 19

Age-length keys, 6

Age sequence: in stock density in East Anglian herring fishery, 15

Aggregation, of herring onto patches of *Calanus*, 241

Ailly, 61, 127

Alaska gyral, 47

Albacore: migration, 11; spawning ground in North Pacific, 13; tagging experiment, 42;

transpacific migration, 62; accessibility to capture, 104; invasion during warm period, 223

Alburnus alburnus Linnaeus, 79

Allen curve, 92, 93

Alternation, between Norwegian and Swedish herring periods, 225

Amino acids, muscle, in cod, 65

Ammodytes marinus Raitt, 150

Analytic models, 183

Anchoveta fishery, Peruvian, 142–44, 172, 218, 219

Anchovy, 117, 227

Anguilla anguilla Linnaeus, 27

Anguilla rostrata (Lesueur), 26, 50

Antartic fin whale: simulated changes in stock density, 181

Antifreeze: in cod's blood, 206

Appearance: of *Velella, Ianthina*, and *Physophora* in British waters, 223; of boreal animals, 223

"Arctic city": where fishermen gather on patches of cod, 104

Arcto-Norwegian cod, 49, 67, 77; spawning ground in Vest Fjord, 13, 22; migration circuit, 24; fin-ray count and vertebral sum, 50; skrei (mature cod), 51; migration from Malangen to Vest Fjord, 51; "fish-carrying layer" in Vest Fjord, 52; dispersal of larvae, 53; Hjort's tagging experiment, 53; Dannevig's tagging experiment, 54, 56; Maslov's tagging experiment, 54, 56; Trout's tagging experiment, 54; echo survey in Ba-

JACKET DESIGN BY CAROLINE BECKETT
COMPOSED BY THE NORTH CENTRAL PUBLISHING CO.
ST. PAUL, MINNESOTA
MANUFACTURED BY CUSHING MALLOY, INC.
ANN ARBOR, MICHIGAN
TEXT IS SET IN TIMES ROMAN, DISPLAY LINES IN SOUVENIR

Library of Congress Cataloging in Publication Data
Cushing, D   H
Fisheries biology.
Bibliography: pp. 259–280
Includes Index.
1. Fish populations.   I.   Title.
QL618.3.C86   1980   597′.05′248   79-5405
ISBN 0-299-08110-9